The
Prolegomena
to a
1985 Philosophiae Naturalis
Principia Mathematica

The
Prolegomena
to a
1985 Philosophiae Naturalis Principia Mathematica
which will be able to present itself as a
science of the true

by

Filmer Stuart Cuckow Northrop, B.A., M.A., Ph.D., LL.D.,
Sterling Professor Emeritus of the Philosophy of Science in the
Department of Philosophy and the Science of Jurisprudence in the
Faculty of Law of Yale University

Together with a Portrait of the Author by Deane Kellar,
Courtesy of the Yale University Art Gallery

The forthcoming volumes are:

Volume I   *1985 Philosophiae Naturalis Principia Mathematica*

Volume II   *et Aesthetica*

Volume III *et Jurisprudencia*

Volume IV   *et Summa Theologica*

Ox Bow Press
Woodbridge, Connecticut

OX BOW PRESS
P.O. Box 4045
Woodbridge, CT 06525

Northrop, Filmer Stuart Cuckow, 1893-
    The prolegomena to a 1985 philosophiae naturalis principia mathematica.

    Bibliography: p. 63
    Includes index.
    1. Science—Philosophy. 2. Mathematical physics. 3. Physics—Philosophy.
4. Knowledge, Theory of. 5. Einstein, Albert, 1879-1955. I. Title.
Q175.N598   1984   501   84-27350   ISBN 0-918024-35-8

Printed in the United States of America

To M.H.C.N.

Table of Contents

Foreword

To understand Professor Northrop's writing in this book, as well as the thinking of those who created both the Greek versions of the Old and the New Testaments, one must realize that the Greeks by their research brought together in Euclid's *Elements* had discovered a multifaceted stable structure which could be expressed axiomatically.

When one starts writing the axioms of such a system that constitute its first principles, one must start with undefined symbols which gain their meaning exclusively from the many-termed relation created by those axioms. Nine undefined terms seem necessary to express the structure discovered by the Greeks: point, line, plane, motion, position, and four kinds of numbers.

For any collection of axioms constituting first principles, Professor Northrop pays respect to the usage of the writers of the Greek versions of both the Old and the New Testaments and uses:

'EN APXAI

which he defines as "first principles." This avoids the usual translations as "in beginning" or "in the beginning" which introduce a concept of time that does not exist as one of the nine undefined terms necessary and sufficient to express the axioms creating the multifaceted stable structure discovered by the Greeks.

Horace G. Oliver, Jr.

Preface

Without any doubt, the two wisest scientific and humanistic minds in our twentieth century are Alfred North Whitehead in the Anglo-American world and Albert Einstein in the Continental European cultured world, with Werner Heisenberg and Erwin Schrödinger coming in a very close third. Stated briefly, the aim of this *Prolegomena* is to do for Einstein's, Schrödinger's and Heisenberg's Equations and their epistemology, with respect to their implications for every one of the humanities (i.e., *Psychologia, Metaphysica, Aesthetica, Jurisprudencia* and a *Summa Theologica*), what Whitehead did for his Equations and their epistemology in his (1) *The Mathematical Concepts of the Physical World* (1906) and *The Concept of Nature* with its epistemological "Method of Extensive Abstraction" (1921) and for their implications in (2) the metaphysics of his monumental *Process and Reality* (Gifford Lectures) (1929), *Religion in the Making* (Lowell Lectures) (1926), *Symbolism and Its Meaning* (University of Virginia Lectures) and his *Adventures of Ideas* (1933) on sociology, cosmology and comparative world civilizations. (He did not write on jurisprudence.)

Unfortunately, there are difficulties in specifying the entailed revolutionary implications for the humanities in the case of the Einsteinean epistemology which do not occur in the case of Whitehead's equations and his less speculative "Method of Extensive Abstraction" epistemology. In Whitehead's case, both his equations and their epistemology were written and published in the English language. Thus, no difficulties in translating from a foreign language into English ordinary language (providing one follows the cautions he gives us) occur in specifying the humanistic implications of Whitehead's equations and his less speculative epistemology. However, in the cases of Einstein's equations and also the Indeterminacy Principle Quantum Mechanical Equation, the difficulties are at least two-fold and quite different.

The first one is that the Einsteinean-Schrödinger-Heisenbergean epistemology is much more complex and difficult to warrant, being much more speculative with respect to its elementary *First Principles* than is the less speculative Whiteheadean "Method of Extensive Abstraction" way of knowing.

The second difficulty is equally serious and less simple to overcome. It arises from the fact that all of Einstein's three major scientific theories — that of his *Special Equation* (1905), his paper on "The Photoelectric Effect" (1917) and his most important paper on the *General Theory of Gravitation* (1916) — and most (though not all) of the publications on *The Indeterminacy Quantum Mechanical First Principle* were written in the German language, the complex clauses-within-clauses syntax of which is that of the single entity-property syntax and its way of thinking, which is true of any ordinary language such as English or even of Newton's Latin (1686). The single paragraphed first page (page 769) of Einstein's epoch-making and greatest *Annalen der Physik Band 49* (1916) original German paper of some 53 pages will suffice to illustrate what the ordinary language and its translation difficulties are.

In his first German sentence on its first page (769), Einstein tells us what he is going to do — namely, generalize his previous *Special Theory* (1905). He then proceeds to tells us with what "pure mathematical aids" (the absolute differential calculus made evident by the investigations of Gauss, Riemann and Christoffel concerning non-Euclidean magnitudes) and with the deductive-inductive mathematical physical aids brought into a system for the solution of physical problems by Ricci and Levi-Civita he (Einstein), in the following 52 pages of his German paper, is going to achieve the generalization of his *Special Theory* (1905).

The *First Principles* (Grundlagen) necessary to do this he achieves on the third page (772) and the seventh page (776) of his paper. From these two revolutionary *First Principles,* written entirely in the German language and *italicized by him,* by way of emphasizing the *First Principles' irreducibility,* he then proceeds to deduce (using the pure mathematical and mathematical physical aids specified) the 75 Equations as Proved Propositions, that is, as operational definitionally prescribed deduced Theorems.

Before we can even begin an ordinary translation of Einstein's 53-page *Annalen der Physik Band 49* (1916) paper from the German, we must realize that neither a single German sentence nor equation can be put into English piecemeal, sentence by sentence, or equation by equation. Instead, we must realize that the two German indicative-sentenced *First Principles* and their warrants must be understood first and only then be translated into English by one who is not misled by the single entity-property syntax and its way of thinking in ordinary

English linguistic prose. Given this English translation correctly, we must realize that all the equations (1-75) on page 769 through 822 of Einstein's paper (1916) follow as proved propositions, i.e., rigorously deduced theorems. This tells us that no sentence or equation in his entire 53-paged paper stands alone. Instead, we are confronted with a deductive-inductive systematic treatise such as Newton's *Philosophiae Naturalis Principia Mathematica* (1686) and the thirteen Greek 'ΕΝ ΑΡΧΑΙ *(First Principles) Books* of Euclid's *Elements* in the fourth century B.C.

To appreciate the English translational difficulties presented by such a deductive-inductive architectonically unified theory of both the natural sciences and of the humanities, it must be realized that our present educational system, from kindergarten through high school and college or university Physics departments or Engineering schools, does not give any training whatever in the pedagogical skills required to understand, to say nothing about translating, from Euclid's Greek, Newton's Latin or Einstein's German into the ordinary English language the complex *First Principles* of Einstein's *Annalen der Physik Band 49* (1916) original German paper and the proved equations those *First Principles* entail.

Nowhere does the remarkable "Economy of Thought" and inductive factual comprehensiveness of this deductive-inductive way of thinking and translating exhibit itself more than in Einstein's dramatic climactic proof of his final equation (75) on the last page (822) of his original German paper. To see why, we have but to note two apparently disconnected scientific and historical events. The first one is that a French theoretical mathematical physicist, in 1843, became famous when, in a paper presented to the French Academy of the Sciences, he proved by an examination of all the optical observational astronomical data from the ancient Greek Hipparchos through Tycho Brahe that the equation for the gravitational constant, later named after this Frenchman the Leverrier Equation, entails a rotational motion of the longitudinal elliptical orbit of the planet Mercury of a factually observed but exceptionally small amount. This was the one and only fact in the entire history of modern science for which Newton's Four Laws of Gravitational Mechanical Motions (1686) could not account. Albert Einstein's deduction from his two German-sentenced *First Principles* established by him in the first seven pages of his *Band 49* of his equation (75) is Leverrier's 1843 Equation for the gravitational constant of Mercury's longitudinal elliptical orbit's motion. In this achievement, the genius of Albert Einstein as a theoretical mathematical physicist (who never performed a physical experiment in his life) exhibits itself.

Clearly, this deductive-inductive way of thinking with its relational n-greater-than-1-adic relational analytic R entitied way of thinking — after the manner of Euclid's 'EN APXAI (English, *First Principles*) (300 B.C.), Newton's *Philosophiae Naturalis Principia Mathematica* (1686) and Einstein's *Die Grundlage zur allgemeinen Relativitätstheorie* (1916) (the *First Principles* of the *General Theory of Relativity*)—we must master in everything which follows in this our present *Prolegomena* and also in all its implications for the humanities in our forthcoming Volumes I, II, III, and IV that are to follow. For this contribution of the theoretical mathematic laws to our task as *First Principles*-focusing philosophers, we must also be grateful.

In the Ox Bow Press reprint (1979) of his *Science and First Principles* (1931), the present writer put this difference between (1) the theoretical mathematician and the philosophers of such exact scientist and (2) the experimental science and the mathematical physical inquirer as follows:

> Science proceeds in two opposite directions from its many technical discoveries. It moves forward with the aid of exact mathematical formulation to new applications, and backward with the aid of careful logical analysis to first principles. The fruit of the first movement is applied science; that of the second, theoretical science. When this movement toward theoretical science is carried through for all branches of science we come to first principles and have philosophy.

This means that there is no such thing as the philosopher of exact science or the humanist coming in *ex cathedra* from outside to tell the theoretical mathematical physicist what his *First Principles* and their *deduced Equations* "really mean". The reason being that because the *First Principles* of the theoretical mathematical physicist are relational analytic R entitiedly defined, the *First Principles* themselves *ipso facto* define that particular theory's meaning. In theoretical mathematical physics, this is what the words *philosophia naturalis* mean.

The theoretical mathematical physicist Albert Einstein gave expression to this irreducibility of the *First Principles* on pages 772 and 776 of his epoch-making *Annalen der Physik Band 49* (1916) paper by putting his first two *Philosophical First Principles* there, in the

German language, *in italics*. Furthermore, these are the only sentences in his original German 53-page paper that are italicized.

Although Alfred North Whitehead's equations and their less speculative "Method of Extensive Abstraction" epistemology are different from those of Albert Einstein, it was nevertheless Whitehead in our first meeting in his Research Office in the Imperial College of Science and Technology in South Kensington's University of London, in September of 1922, who first made the present writer aware of the cause of not taking the theoretical mathematical physicists' verified-by-others *First Principles* as irreducible. His caution then was:

> We cannot be too suspicious of ordinary language in science and philosophy, the reason being its single entity-property syntax.

How to implement this warning in the case of the different equations and their more speculative philosophical *First Principles* of Einstein and also of the Indeterminacy Principle Quantum Physicists must and shall concern us in this *Prolegomena*'s sections S9 through S12. However, "the single entity-property syntax" factor in Whitehead's final clause just above hints at the way, e.g., "the disembodied mental substances" of Descartes that could think but not feel; those of Locke that could sense but not think; and the embodied "material substances" of Hobbes, the materialistic part of Marx, and the present-day Harvard behavioristic psychologist, Professor B.F. Skinner, whose bodily persons can neither think nor feel; together with the pseudo "body-mind problem" and other pseudo-notions must all be eliminated. They point up the practical direction which our thinking and acting must take if these pseudo-entities and the pseudo "body-mind problem" and other paradoxes that they generate are to be avoided.

This direction was pointed out also by Whitehead on that September afternoon in 1922 when he added:

> There is no solution to the problems of modern science and philosophy at our present twentieth century end. A mistake was made in the interpretation of the *First Principles* of Galilei, Descartes, and Newton's Physics at its founding, and we must first correct that initial error, and then go on afresh with wider general theory from there.

In this connection, we need to follow the pedagogical warning of New England's beloved philosopher, the late John Dewey, who in his Gifford Lectures on the danger of *The Quest for Certainty* wrote to the following effect:

In our pedagogical practices from Kindergarten through graduate and professional schools we need to give as much attention to first getting the old outmoded notions out of the learner's mind as we give to putting the new and more factual comprehensive ones in.

In any event, because the philosophical *First Principles* in the speculatively discovered verified-by-others way of human knowing in the Einsteinean epistemology is a much more complex one than that of any previous theory, we must expect the task of specifying its humanistic implications in this our present *Prolegomena* and its four much larger future volumes to be a prodigious one. Even so, here again Alfred North Whitehead, in a footnote on page 508 of his monumental *Process and Reality* (1929), expresses the following need for its achievement and the likelihood of success:

> The theory of the derivation of the basic uniformity requisite for congruence and hence for measurement should be compared with that by Professor F.S.C. Northrop of Yale. I cannot adjust his doctrine of a "macroscopic atom" to my cosmological outlook. Nor does this notion seem necessary if my doctrine of "microscopic atomic occasions" be accepted. But Professor Northrop's theory does seem to be the only alternative if this doctrine be abandoned.

In any event, our four forthcoming volumes are:

Volume I   *1985 Philosophiae Naturalis Principia Mathematica*
Volume II   *et Aesthetica*
Volume III   *et Jurisprudencia*
Volume IV   *et Summa Theologica*

A word about our *Prolegomena*'s subtitle: the word *Prolegomena* itself, like its subtitle, is from Kant, but written in English, rather than in Kant's Latin of 1783, the reason being that we follow both Whitehead and Einstein in correcting the errors made in the interpretation of Galilei, Descartes and Newton's physics by returning today's *Philosophia Naturalis* to the verified *'EN APXAI (First Principles)* of the ancient Greeks, as given to us in the thirteen Greek books of Euclid's *Elements*. Our present crucial experimentally verified *Epistemological First Principles* are further and, as our four future volumes will show, richer, natural and humanistic generalizations of them.

F. S. C. Northrop
151 Water Street
Exeter, N.H. 03833

An Introduction:
Contemporary Einsteinean Epistemology
and
Neoplatonic Christianity

This *Prolegomena* was first read at the request of Professor R. Baine Harris, President of the International Society for the Study of Neo-platonism, at this Society's Sunday Morning Meeting, held at the Catholic University of America, in Washington, D.C., on October 15, 1978. The theme of the meeting was "Neoplatonism and Christianity." This makes it appropriate on this Sabbath morning to preface what follows with two texts, each by a professional expert who is also a committed Christian. Our newer Testament one — lest our religion lack relevance and utility — is by today's feminine economic scientist of utility, Barbara Ward. She writes, "Faith will not be restored in the West because people regard it to be useful. It will return only when they find it to be true." In short, a viable Christianity must be a cognitive one, found to be scientifically true by scientific experts who convey this truth to people generally. Our older Testament text — lest we ignore what the words "Neoplatonism" and "Christianity" mean — is by the late Dean Inge, the English-speaking world's leading authority on Plotinos, who was the creator of Neoplatonic Christianity. In the late 1920's, the Dean of St. Paul's said, "The religious and philosophical implications of recent advances in science are . . . the battleground on which the defense of Christianity against its enemies must be fought out."[1] He also added[2] that those who would follow Luther and Harnach's inclination to by-pass Greek Christian civilization for nothing but its Hebrew component risk destroying Christianity itself. In his *Untergang des Abendlandes*,[3] Oswald Spengler made the same warning for Western science and civilization generally.

---

[1] William Ralph Inge (Gifford Lecturer), *The Philosophy of Plotinos*, Volumes I and II (1916-1917) (published 1923), Longmans, Green & Co., London, Boston, Madras & Calcutta.

[2] William Ralph Inge (Terry Lecturer), Yale University (1924), Yale University Press, New Haven, Conn.

[3] Oswald Spengler, Der Untergang des Abendlandes, Volumes I and II (1923), C.H. Beck München.

Einstein's role in "these advances in recent science" is well known. Not so familiar is his conviction that Euclid's Geometry must become our model if today's verified theoretical physics is to be understood.

S2
Einstein on Euclid's Geometry

In a statement published in 1934 in his scientific paper on his special branch of today's science, "The Method of Theoretical Physics," Einstein writes:

> We reverence ancient Greece as the cradle of Western science. Here for the first time the world witnessed the miracle of a logical system which proceeded from step to step with such precision that every one of its deduced propositions was absolutely indubitable — I refer to Euclid's Geometry. This admirable triumph of reasoning gave the human intellect the necessary confidence in itself for its subsequent achievements. If Euclid failed to kindle your youthful enthusiasm, then you were not born to be a scientific thinker.[4]

This means that the Greek defined Axioms, Postulates, Common Notions, and their deduced Proved Propositions, as given to us in *T.L. Heath's Three Volumes* (1905) H-I, H-II and H-III of the Heiberg Greek Text of Heath's edition of the *Thirteen Books of Euclid's Elements*,[5] must be one of our bibles also. To be noted also is the fact that the *First Principles* of verified ancient Greek science were all defined in the Greek Definitions of this Heath-Heiberg Greek text; and the Greek name for them in all Greek physics, metaphysics and theology is 'EN APXAI (English, *First Principles*),[6] which means "Elements." For reasons which will become evident later, no English ordinary language translations are to be ventured by the prudent reader, only Heath's English translation of the Heiberg Greek text is to be used or trusted, the reason being that only he knew the ancient and

---

[4] Albert Einstein in *The World As I See It* (1934), pages 31-32.

[5] T.L. Heath, *The Thirteen Books of Euclid's Elements* (1908). Translated from the Greek text of Heiberg with his Introduction and Commentary: Volumes I, II and III, hereafter (H-I), (H-II), (H-III), with Greek-English Lexicon of Technical Scientific Terms at the end of each Heath Volume (H-I), (H-II) and (H-III). See (H-I) on Euclid's Postulate 5 on pages 202-220. In the Index of each Heath volume will be found the names, dates and achievements of every Greek and Greek-into-Latin ancient Greek scientist, philosopher and theologian from Anaximander in 610 B.C. through the Neoplatonists up to 660 A.D.

[6] Witness in Heath's Section S3 on page 117 of his (1908) Volume H-I his expanded Title for the English Elements: "*First Principles: Definitions, Postulates, and Axioms.*

modern verified 'EN APXAI mathematical physics, as well as his classical Greek and Latin.

Heath's Greek Definitions are essential in our entire inquiry, as are his Commentaries and his Greek-English Lexicon on them, for a second most important reason. They provide the connecting link for relating the epistemological way of knowing in Einstein's verified Theories of Relativity and in Indeterminacy Principle Quantum Mechanics both to the later Neoplatonism and also to Greek-into-Latin Christianity. The relation between the 'EN APXAI of Euclid's "Elements" and those of the Greek and Greek-into-Latin First Principles of Neoplatonism is well known to all Neoplatonic scholars.[7] The converse influence is also true, as the Commentaries of Proclos on the Parallel Postulate (5) in Euclid's *Book I* show. To evaluate the validity of Proclos' and other Neoplatonists' criticism of Euclid because he did not prove this proposition as a rigorously deduced theorem will be imperative in our forthcoming Volume I (*Philosophiae Naturalis Principia Mathematica*) and for this we must read both the Greek texts as given in Heath's (1908) *Volume H-1* and his lengthy Commentaries which embrace not merely the ancient Greek but also the modern mathematical physicists, such as Bolyai, Lobachewsky, Riemann, Barrow, Newton and the more recent very important De Morgan.

Not so well known is the fact that all Hebrew-into-Septuagint Greek and Greek New Testament scholars, such as Philo of Alexandria (c. 30 B.C.-45 A.D), Tertullian (c. 160 A.D.-240 A.D.), Origen of Alexandria (185 A.D.-254 A.D.), Clement of Alexandria (150 A.D.-216 A.D.), Ptolemaios Klaudios' *Syntaxis Mathematica* (140 A.D.) (Arabic, *Almagest*), Proclos, Plotinos (205 A.D.-270 A.D.) and the great Greek scientifically informed, Latin-writing St. Augustine (354 A.D.-430 A.D.), regarded the Hebrew-into-Septuagint Greek-Greek New Testament 'EN APXAI to be true because their Definitions were verified as given to us in Euclid's *Thirteen Books*. The demonstration of this will concern us in the sequel, and more fully, in our future Volume IV: *et Summa Theologica*.

Of equal importance is Einstein's use of the adjective "theoretical" in his 1934 paper's title referring to Euclid. This adjective

---

[7] Glenn R. Morrow, *Proclus: A Commentary on the First Book of Euclid's Elements*, translated with *Introduction*, pages vii-xliv and *The Commentary*, pages 3-345. This treatise must be supplemented by the Greek and Greek-into-Latin text of Professor B.L. van der Waerden (1966 and 1979) and Professor Samuel Sambursky (1956, 1959 and 1962) referred to in the later sections (S13 and S14) of this *Prolegomena*.

"theoretical" bespeaks the fact that although he made several hypothetical, speculative, imaginative experiments as described in his scientific papers, he never performed an actual physical experiment in his entire life. Instead, like Euclid, Parmenides, Zeno, Anaxagoras, Plato and Theaetetos in ancient Greek verified science; Maxwell and Willard Gibbs in recent times; and Planck, Schrödinger, Heisenberg, Bohr, Pauli and Dirac in our own remarkable century, Einstein's genius is that of the theoretical, not that of the experimental, natural scientist, important and necessary though the latter be for the verifications of the theoretician's theories. This makes the truth-value of their achievements all the more impressive, since it necessitates that the verifications be made by mathematical minded experimental physicists such as Arthur Compton, Bridgman,[8] and Davisson and Germer, rather than by the discoverers and constructors of the theories themselves.

S3

The Method of Theoretical Mathematical Physics

The late philosopher of exact science Professor Morris R. Cohen, who visited Einstein many times after the latter moved from the University of Berlin to the Princeton Institute, gave the present writer the following account of Einstein's amusement over the statements on the front page of *The New York Times* and elsewhere to the effect that he was "the world's greatest experimental scientist." Einstein said that upon one occasion he thought he could test on the opposite wall of his study the change in the size of the projected shadow of a lighted candle as he walked toward his study's wall, whereupon Einstein then related: "You know, the hot wax burnt my thumb. I dropped the candle and never did find out how that experiment ended."

For this major role of the theoretical mathematician in discovering the theoretical *First Principles* of today's and yesterday's exact verified science while leaving to the experimental physicists their verifications, we must be grateful. In the Preface to our *Science and*

---

[8] Percy W. Bridgman, *The Logic of Physics* (1924), Macmillan, N.Y.
[9] Morris H. Cohen, *Reason and Nature* (1931), Harcourt Brace & Co., N.Y.

First Principles (1931),[10] the present writer put the reason for this gratitude as follows:

> Science proceeds in two opposite directions from its many technical discoveries. It moves forward with the aid of exact mathematical formulation to new applications, and backward with the aid of careful logical analysis to first principles. The fruit of the first movement is applied science; that of the second, theoretical science. When this movement toward theoretical science is carried through for all branches of science we come to first principles and have philosophy.

No one has given a better description, together with the operation definitional methods by which it is crucially experimentally verified, of this Einsteinean epistemological way of knowing than did Henry M. Sheffer in 1921 in his course on Mathematical Physical Logic in which he defined *Deduction* as *the search for Minimal Logical Universal Antecedents* and *Induction* as *the pursuit from them of Maximal Logical Consequents.*[11] In his 1922 research seminar, composed of but four students, of whom the present writer was one, Sheffer went on, in his *Relational Logic of Systems and System Functions,*[12] to specify, both analytically and with graphical matrices, the postulational techniques by means of which n greater than 1-adic formal relational properties are built into a relational analytic entitled relation R to define thereby all the entities in that specific R's field.

The formal equivalent of this is the Columbia University mathematician Cassius Keyser's *Relational Logic of Doctrines and Doctrinal Functions,* as given in his *Mathematical Philosophy.*[13] The present writer has used the texts of both Sheffer of Harvard and Keyser of Columbia in his teaching at Yale University over a period of 39 continuous years and found that the relational terminology of Keyser's *Relational Logic of Doctrines and Doctrinal Functions* fits the religious and humanistic concerns of our overall inquiry. Also, because it derives from the

---

[10] F.S.C. Northrop, *Science and First Principles* (1931). Hereafter NSFP. *Preface.* Cambridge at the University Press, England. Macmillan, N.Y. 1979 Reprint. Ox Bow Press (1979), Woodbridge, Conn.

[11] Henry M. Sheffer, *A General Theory of Notational Relativity.* Printed booklet on *Empirical and Relational Logic* (1921). Harvard University Press, Cambridge, Mass.

[12] Henry M. Sheffer, *Relational Logic of Systems and System Functions with Analytic Postulational Techniques and Diagrammatic Matrices* (mimeograph) (1921), Cambridge, Mass.

[13] Cassius Keyser, *Mathematical Philosophy* (1916), Dutton Publishers, New York; Louis Rougier, *La Structure des Théories Déductives* (1921), Alcom, Paris; and Ian Mueller, *Philosophy of Mathematics and Deductive Structure in Euclid's Elements* (1981), M.I.T. Press, Cambridge, Mass., U.S.A. & London, England.

Dedekind, Riemann and Hilbert relational theory of pure mathematical entities and *ipso facto* of mathematical physical ones, it prepares the introductory student for what he finds in his Mathematics, Physics and Engineering departmental courses and also prevents him from supposing that relational analytic R defined mathematical physical entities can be defined by the extensional logical calculus alone.

Even more important on the philosophical *First Principles* side, it prevents elementary students and everyone else from falling into the single entity-property notions of the $\phi(x)$, $\supset$, $\psi(x)$ (material implication extensional logical calculus) of "The Theory of pure mathematical Deduction" in the Whitehead and Russell's *Principia Mathematica* (1910)[14] which (though very important in Cybernetic Neural Net Firings [positive and negative feed-back] Theory, as the sequel will show) is quite inadequate as a sufficient theory of mathematical-physical n greater than 1-adic R defined scientific entities as Whitehead made evident in his *Concept of Nature* in 1922[15] and as the Harvard Gibbsean-minded physiological chemist L.J. Henderson made evident to the present writer in 1921.[16] The reason why such is the case, as our sequel will demonstrate, is that all modern and present-day verified theoretical mathematical physical exact science rests on Galileo's *Principle of Relativity*, and what this principle affirms is that all veridical scientific entities are n greater than 1-adic relational analytic R defined and not single entity-property ordinary language "mental and material substances" ones. Einstein gives expression to this Galilean relational theory of all entities First Principle when he writes:

> In a consistent theory of relativity there can be no inertia *relative to 'space'*, but only an inertia of matter relative to one another.[17]

Unfortunately, most traditional modern philosophers and psychologists have not been governed by this Principle of Relativity, which it was the genius of Galilei to have discovered first in the founding of modern science. This forces us to face the following topic:

---

[14] Whitehead and Russell, *Principia Mathematica* (1910), "The Theory of Deduction," Introduction, page xv, University Press, Cambridge, England.

[15] Alfred North Whitehead, *The Concept of Nature* (1920), Chapters VI and VII on "Recognition" and "Revelation," University Press, Cambridge, England.

[16] L.J. Henderson, *Nomogram of Gibbsean Steady State in Human Blood* (1921), published in 1931 in Chapter IV of F.S.C. Northrop's *Science and First Principles* (1931), University Press, Cambridge, England; Macmillan, N.Y.

[17] As quoted in NSFP, page 83.

S4

Some Tragic Effects of the Extensional Logical Calculus and of Single Entity-Property Thinking in General

The erroneous non sequitur that (1) the $\phi(x)$, $\supset$, $\psi(x)$ extensional logical calculus is all-sufficient in the sense of building for Principle of Relativity mathematical physical entities as well as for the Whitehead-Russell (1910) pure mathematical one is the error of all Vienna positivists of the Schlick-Carnap school. Even Whitehead, who held this theory for pure mathematical entities, rejected it in his theory of "recollection" and of memory, affirming in the middle chapters of his *The Concept of Nature* (1921) that "the ingression of sensa into nature is in many termed relation" for mathematical physical entities.

Equally erroneous are the single entity-property entities of (1) *the mental substances-only* disembodied *Geiste* (English, *Ghosts*) of both the pluralistic Bishop Berkeley, Leibnitz' "windowless monads" and the *categorical a priori* "Forms of Sensibility" — namely, *Euclidean absolute metrical space* and Newton's *one dimensional absolute time flow* of Kant's *disembodied Transcendental Ego* in the latter's *First Critique* (1871).[18] The same holds for (2) the *Hypothetical a priori Transcendental Ego* of the early nineteenth century's Cohen of Marburg, *Neo-Kantian*, and also for the present-day *London School of Economics Neo-Kantian* Professor Karl Popper.

Nor does the monistic Hegel's correct observation that there then is no meaning for these pluralistic "Idealist's" public world, except as the Kantian or Neo-Kantian Forms of Sensibility are identical in all knowers. This entails, as Marx, Royce and the English Bradley following Hegel saw, that the Kantian and Neo-Kantian disembodied *Geiste* are not many human *Ghosts,* but instead are *one absolute transcendental Egotistical Ghost,* their introspected plurality being completely phenomenal and phantasmically non-objective and hence *a posteriori.* Equally phantasmic and fantastic, at the other extreme, are the embodied feelingless and thoughtless single entity-property material persons of the materialistic Hobbes, the materialistic "Religion is the opium of the people" atheistic piece of Marx, and also the present-day behavioristic Professor B.F. Skinner of Harvard Univer-

---

[18] Immanuel Kant, *Kritik der reinen Vernunft* (1781).

7

sity's Department of Psychology, who believes that he himself and also you and I have no capacity to feel or think, both feeling-awareness and thought being phantasmically phenomenal and non-existent.

Nor were matters helped the slightest when a few years ago in a joint public announcement Professor Karl Popper, who believes himself to be a Neo-Kantian *Ghost,* and Sir John Eccles, who as a Dualist believes himself to be a *different Neo-Kantian Ghost* (than Professor Popper) in interaction with the countless single entity-property feelingless and thoughtless material atoms of his (Sir John's) body, proclaimed that in a forthcoming research investigation, to be held in romantic Lake Como's Castle Benelli, they would produce their "Solution of the Body-Mind Problem." Overlooking the fact that their common single entity-property way of thinking is the cause of their disembodied Ghostly-embodied Ghastly materialistic notions, they both disregard the latent fact that, single entity-property thinking being the source of their initial predicament, both disembodied Ghostly minds and feelingless and thoughtless Ghastly material substances being phantasmic non-objective pseudo-entities, their "body-mind problem" is a "pseudo problem" also, there being no "body-mind problem" to solve in the first place, or disembodied Ghostly minds, or feelingless and thoughtless human persons either.

In thus exorcising these disembodied *Ghosts* and feelingless and thoughtless *Ghasts* (which arose with Descartes, Hobbes, the early naive realistic "blank tabletish" Locke and was carried on by Berkeley, Leibnitz with his claustrophobically-trapped privately introspecting "windowlessly monadic persons", and by Kant, Hegel and the Neo-Kantians), we *correct the misinterpretations* of the *First Principles of Modern Science,* which Whitehead noted in 1921, and we are at last able to return to our positive task. It is:

S5

The Scientific Method of Natural History Biology, Medicine and Psychology

The most famous and remarkable recent representative is the giant Darwin, as exemplified in his *The Voyage of the Beagle* (1836) and *The Origin of Species* (1859). Its founder was, however, the great ancient fourth-century B.C. natural history biologist, Aristotle (384 B.C.-

322 B.C.), whose *Logic of Predictables*[19] defined the complete inductive descriptive and classificatory method of this verified natural history science as given in his two classic treatises, *Historia Animalium* and *De Partibus Animalium.* Without them, by way of the medieval Paracelsus and the modern Buffon, Cuvier, Linnaeus and Wallace, Darwin's *Voyage of the Beagle* (1836) and *Origin of Species* (1859) would not be.

Unfortunately, due to the erroneous single entity-property notion of physical objects described in our previous section, many scientists who should know better and even more humanists have read these non-existent pseudo entities into the "chance variational" words of Darwin's *Origin of Species,* thereby turning Darwin into a pseudo materialistic metaphysician and discrediting Aristotle also. Whereas, the fact is that both of them share the same exceedingly important method that is common to Natural History Biology, Medicine and Psychology.

To understand why, it is imperative that no one should read any of Aristotle's later works, *Physica, Metaphysica, De Anima* or *Theologica,* unless he has first read Aristotle's earlier two natural history biological treatises, *Historia Animalium* and *De Partibus Animalium.* Why? The reason is that a reading of the two latter treatises shows that not a single category of Aristotle's *Metaphysica* and other later books appears in them. Instead, his *Historia Animalium* and *De Partibus Animalium* restrict themselves entirely to meticulously describing the directly sensed properties of all the diverse species of animals and of their parts also. Thus it is that Aristotle is the founder of Natural History Descriptive Zoology, of which Darwin's treatises are a nineteenth-century example.

Even so, there is something exceedingly important that every one of us can and must do if we are not (as practically everyone today has done) to misinterpret not merely Aristotle but also the great Darwin's *Voyage of the Beagle* and *The Origin of Species.* The reason is that Aristotle's *Historia Animalium* and *De Partibus Animalium,* being natural history initial stage of scientific inquiry works, stand or fall on their own feet, quite independently of Aristotle's *Physica, De Caelo* (astronomy) and *Metaphysica.* Hence, the latter can be partial or, as is the case in the physics, false, with the *Historia Animalium* and the *De*

---

[19] Aristotle, *Historia Animalium* and *De Partibus Animalium,* English edition, ed. W.D. Ross, Oxford University Press.

Partibus Animalium still as true today as it was in the fourth century B.C. when Aristotle made the completely inductive meticulous description of actual animals, their parts, and their history which makes up their contents. This is why as the founders of the science of Natural History Biology, Galen, Paracelsus, Cuvier of the Jardin des Plantes, Linnaeus, Wallace and Darwin of both *The Voyage of the Beagle* and *The Origin of Species* stand on the great Aristotle's shoulders.

Nothing is more erroneous than the omniprevalent present-day notion that Aristotle's natural history biology has been contradicted by the theory of evolution of Darwin's remarkable treatises. His and Aristotle's biological works are inductive descriptive genus-species classificatory natural sciences. Such sciences, as the present writer has noted in his chapters on them in his *The Logic of the Sciences and the Humanities*,[20] are strong and endlessly voluminous on descriptive power, weak on predictive power, and with no "Economy of Thought" whatever.

Furthermore, Aristotle would welcome the evolution of new species in time as well as in space, as the word *Historia* in his *Historia Animalium* demonstrates. Also, because his two natural history biological works stand on their own feet, they do not entail the teleological causality of his doctrine of all objects "seeking their natural places" of his *Physica, Metaphysica* and *Theologica*, which does get one into a contradiction with not merely Darwin's "chance variational" *Origin of Species* but also with the crucial experimentally verified mechanical causality of Galilei's and Newton's mathematical theories and also of Einstein's Theories of Relativity (1905, 1916) and the Indeterminacy Principle Quantum Mechanics (1900-1931).

There is also a converse error, omnipresent in today's Judaic and Judaeo-Christian theological literature, which the *Historia Animalium* and *De Partibus Animalium* independent interpretation by St. Thomas Aquinas of Aristotle's Physics and Metaphysics avoids. This error arises from the error noted above, of reading the single entity-property feelingless and thoughtless *Ghasts* into Darwin's "chance variations" and also into the theoretical mathematical physical crucially experimentally verified entities of all of natural sciences, in their implications for the humanities, thereby turning countless initially believing Jews and Christians, after the manner of the materi-

---

[20] F.S.C. Northrop, *The Logic of the Sciences and the Humanities*, Macmillan, New York (1947); Reprint 1983, Ox Bow Press, Woodbridge, Conn.

alistic piece of Karl Marx, into at least agnostics if not, as with him, outright atheists, believers that "Religion is the opium of the people." This is also one reason why this *Prolegomena* must be followed by its forthcoming Volume IV, entitled *et Summa Theologica*.

So much for the initial natural history stage of scientific inquiry in biology with its "sense world contains descriptively the real world" epistemology, as compared with the much more complex way of knowing with its remarkable mechanical causal predictive power of the theoretical dynamics theories of Einstein's Theories of Relativity (1905, 1916) and the Indeterminacy Principle Quantum Mechanics (1900-1931). If the different usages of the "cause" in these quite different sciences are not to be confused, with each corrupted, it is necessary that we now analyze and define a science that its experts call a "theoretical statics" in which its use of the word "cause" is stronger than that of any natural history inductive descriptive science but very much weaker than the Einsteinean "theoretical dynamics" use of the word "cause."

The middle ground species which has a "theoretical statics" but no "theoretical dynamics" we shall denote as THS. An example which we shall analyze and define is the Neo-classical Austrian-English Marshallean Economic science. It can be best approached by way of an autobiographical experience. In the early 1940's, the present writer was introduced to the English Marshallean economist in the London School of Economics, Professor Lionel Robbins. He said:

> You have been investigating and publishing analyses of the rigorous methods of the theoretical mathematical physical sciences. Why don't you also turn your attention to our science of Economics? I am convinced that we achieve a predictive power greater than that of your American Wesley Mitchell's extrapolation of leading, following and lagging indicator inductive curves, but not as powerful as that of the exact theoretical mathematical physicists.

A few months later, Professor Schumpeter of the Austrian Neo-classical School was giving a graduate seminar in the Yale University Department of Economics in which he gave evidence that he had a theoretical economic dynamics. Because of Professor Robbins' suggestion and the knowledge of how complex and difficult the achievement of a theoretical dynamics in mathematical physics is, the question was raised as to whether such a (TDS) theory had finally been achieved in economic science. Whereon the present writer was challenged to read a paper on this subject in his graduate seminar but four evenings later. Never was a paper written so quickly.

In this paper, its writer's analysis established Professor Lionel Robbins' claim that economic science has the limited predictive power of a theoretical statics, but not the claim that it could because of its elementary First Principle ever achieve the unqualified mechanical causal predictive power of a (TDS) theoretical dynamics.

With knees knocking and perspiration on his brow, the reader ended, whereupon immediately, without waiting for anyone to raise a question, Professor Schumpeter, in his gracious Viennese manner, turned to the paper's writer and said:

> Will you give me the honor of presenting your paper to the editor of the *Harvard Quarterly Economic Review* for immediate publication in its next issue.[21]

The next week, the expert on the method of economic science, Professor Morgenstern heard about this paper and invited its author to present it at the Economics Club of Princeton University. The result was the same. Afterward privately, he told its reader that the reason Professor Schumpeter thought he had an economic dynamics is that he took for his economic data the ship-building industry, which has a minimum span of seven years between the receipt of firm orders and the building of the finished ship, and also that he had the good fortune to choose the beginning of this seven-year span as the initial state of the economic system being studied. Had the final month of the seven-year span been picked, his deduction of the future state of the system would be wrong within a few weeks.

Professor Lionel Robbins in his classic *The Nature and Significance of Economic Science*,[22] referring to Winston Churchill's experience at the end of World War I, makes clear why Professor Morgenstern was right, in the following way:

> After years of effort, the nation had acquired a machine for turning out the materials of war in unprecedented quantities. Enormous programmes for production were in every stage of completion. Suddenly the whole position is changed. The 'demand' collapses. The needs of war are at an end. . . . What is relevant is that what at 10:55 a.m. that

---

[21] F.S.C. Northrop, *Quarterly Journal of Economics* (November 1941). Reprinted in the writer's *The Logic of the Sciences and the Humanities*, Chapter XIII, Macmillan, N.Y. (1947), reprint, 1983, Ox Bow Press, Woodbridge, Conn.

[22] Lionel Robbins, *The Nature and Significance of Economic Science* (1935), Macmillan Ltd., London.

morning was wealth and productive power, at 11:05 had become 'not wealth', an embarrassment, a source of social waste. The substance had not changed. The guns were the same. The potentialities of the machines were the same. From the point of view of the technicians everything was the same, but from the point of view of the economist, everything was different. Guns, explosives, lathes, retorts, all had suffered a sea change.

In short, what constitutes an economic want is not the guns and the lathes as objective physical entities but the wanting of them by individual persons. Also, this applies to ships as well as to everything else. Professor Robbins puts it in this way:

> There is no quality in things taken out of their relation to men which makes them economic goods. There is no quality in services taken out of their relation to the ends served which makes them economic. Whether a particular thing or a particular service is an economic good depends entirely on its relation to valuations (i.e., individual human preference).

He then adds vis-à-vis the post hoc profits of the historians:

> The bore at the Club announces, "History proves," and one reconciles oneself to 45 minutes of *non sequiturs*. [23]

Quincy Wright of the Department of Economics in the University of Chicago adds:

> There is nothing in Economics corresponding to momentum or energy, or their conservation laws. [24]

As our Section S6 shows, the latter is what gives the exact theoretical mathematical physical crucially verified sciences their remarkable predictive power.

These considerations make it imperative that we now rigorously define the three following species of science. We shall denote them respectively as (NHSI), (TŞT) and (TDT).

(NHSI) The Natural History Stage of Scientific Inquiry. Its logic is the Logic of Predictables of Aristotle.

(TŞT) Let S be an $n>1$-adic relational analytic R entitiedly defined First Principle science such that, given the empirical values of the independent variables, its state-function at any present moment of time t^0 can be deduced.

---

[23] Lionel Robbins, *The Nature and Significance of Economic Science* (1935), Macmillan, London.

[24] F.S.C. Northrop, published in *Science and First Principles* (1931), University Press, Cambridge, England; The Macmillan Co., N.Y., 1979 Reprint, Ox Bow Press, Woodbridge, Conn. On Einstein's *Special Theory*, see its Chapter II.

(TDT)   Let D be an n>1-adic relational analytic R defined First Principle science such that, given the values of its independent
variables at any present moment of time t^0, then the values of
its independent variables, not merely at any later moment of
time $t^0 + 1$ can be rigorously deduced, but also those of any
past moment of time $t^0 - 1$ can be also.

The time, at long last, has now arrived to give an example of the
last species.

S6

Einstein's Special Theory of Gravitational and Electromagnetic Dynamics (1905)

Of the prodigious "Economy of Thought" which isomorphically
identical *irreducible relational n>1-adic relational theory R of all entities,*
in any specific scientific domain, provides, we as philosophers of such
speculatively discovered verified-by-others scientific philosophy
must take advantage. In the 1979 Preface of the 1979 reprint of his
Sheffer-L.J.Henderson-William Ernst Hocking directed Harvard
Ph.D. Thesis, "The Problem of Organization in Biology" (1924),
which was published in 1931 under the title *Science and First Principles*
(1931), the present writer stated this Economy of Thought contribution by the theoretical mathematician to our *First Principles*
(seeking and describing) philosophical task, as follows:

> The theoretical mathematical scientists separate out for us the differ-
> ence between (a) the method of the theoretical mathematical physicist
> and (b) the method of the philosopher of such exact relational entitied
> R defined science. The important thing to note about the philosopher's
> method (b) is that it involves no speculation on the philosopher's part.
> Speculation is scientifically trustworthy only when practiced by the
> epistemological method (described best by Einstein) by rare theoretical
> mathematical scientists of genius. Hence, the philosopher's task does
> not begin until such a theoretical mathematical physicist presents him
> with a speculatively discovered theory and mathematical physical
> minded experimental physicists assure the philosopher that this theory
> is verified.

These conditions being met, let us now turn to Einstein's Special
Paper (1905). With respect to this remarkable paper, it is imperative
that we use only his German text, the reason being an article published in Volume 31 (1963) of the *American Journal of Physics* by the

Princeton University educated Charles Scribner, Jr., who today is president of the Charles Scribner's Sons Publishing Company in New York City.[25]

Mistranslation of a Passage in Einstein's Original Paper on Relativity

Students of the Special Theory of Relativity who have read Einstein's fundamental paper "Zur Elektrodynamic bewegter Körper" only in the English translation by W. Perrett and G.B. Jeffery may appreciate our pointing out the following passage in which an unnecessary elaboration of the text on the part of the translators has somewhat altered its exact meaning. We quote below the original passage and then the translation, placing brackets around the questionable part of the latter.

"Es ist aber ohne weitere Festsetzung nicht möglich, ein Ereignis in A mit einem Ereignis in B zeitlich zu vergleichen; wir haben bisher nur eine 'A-Zeit' und eine 'B-Zeit,' aber keine für A und B gemeinsame 'Zeit' definiert. Die letztere Zeit kann nur definiert werden, indem man *durch Definition* festsetzt, dass die 'Zeit,' welche das Licht braucht, um von A nach B zu gelangen, gleich ist der 'Zeit,' welche es braucht, um von B nach A zu gelangen."

"But it is not possible without further assumption to compare, in respect of time, an event at A with an event at B. We have so far defined only an 'A time' and a 'B time.' We have not defined a common 'time' for A and B, [for the latter cannot be defined at all unless we establish *by definition*] that the 'time' required by light to travel from A to B equals the 'time' it requires to travel from B to A."

A faithful translation of the last sentence of the passage in question would read simply: "The latter time can now be defined in establishing *by definition* that the 'time' etc."

In his Stafford Little Lectures delivered in May 1921 at Princeton University, Einstein remarked that:

"The theory of relativity is often criticized for giving, without justification, a central theoretical role to the propagation of light, in that it founds the concept of time upon the law of propagation of light. The situation, however, is somewhat as follows. In order to give physical significance to the concept of time, processes of some kind are required which enable relations to be established between different places. It is immaterial what kind of processes one chooses for such a definition of time. It is advantageous, however, for the theory, to choose only those processes concerning which we know something certain. This holds for the propagation of light *in vacuo* in a higher degree than for any other process which could be considered, thanks to the investigations of Maxwell and H.A. Lorentz."

---

[25] Charles Scribner, Jr., *American Journal of Physics*, Vol. 31 (1963).

It would seem possible that the above-mentioned liberty by the translators has helped to perpetuate a misunderstanding of Einstein's own
position on this point.

CHARLES SCRIBNER, JR.

Charles Scribner's Sons
597 Fifth Avenue
New York 17, New York

The German text of Einstein's Special Theory (1905) is derived in
his original German *Annalen der Physik Band 49* (1916), Seite 769-822
(1916) Theory, as a special restricted case. Should some critic ask the
question "How could the 1905 text be derived from a 1916 paper?"
he would be guilty of confusing the temporal relation R of "1916
minus 1905, eleven years later" with the totally different n greater
than 1 adic Logical First Principal R, the remarkable logical universality that Professor Walter Thirring so dramatically describes in a
1972 paper. To this we must now turn.

S7

Professor Walter Thirring's Classic Paper
on Gravitation (1972)

This paper opens with the following sentence: "The problem of gravitation was solved by Einstein in 1916." (Of this, more shortly.) He
goes on to say:

> The more important feature of gravitation is its universality which goes
> much beyond that of other interactions. It appears in the legendary
> version of Newton under the apple tree where he sensed that every-
> thing, we, the apple, the moon are subject to the same universal force.[26]

He then describes its exceptional verified-by-others experimental accuracy in his next page (127):

> In fact, elementary particle physics has provided a fantastically sensi-
> tive instrument which allows us to verify this fundamental property of
> gravity with a precision of at least 1 in 10^{14}.

In short, the chance of Einstein's Minimal Logical Universal
Antecedent First Principle (EGT) being in error, due to human
experimental errors on the experimental physicist's part, is but 1

---

[26] Professor Walter Thirring on *Gravitation* (1972), in Cohn, G.K.T. and Fowler, G.N.
(1972); Chapter 4 in Volume 4 of *Essays in Physics*, Academic Press, London and N.Y., pages
125 ff.

chance in 1 with 14 zeros after it. In everything that follows in this entire volume we can, therefore, take Einstein's Minimal Logical Universal First Principle as being remarkable both in its logical universality and its crucial experimental verification by experimental physicists other than Einstein himself.

S8

The First Principles of Einstein's *Annalen der Physik Band 49* (1916) Paper on Theoretical Dynamic Gravitational Mechanics

Because of the English translational reasons, demonstrated by Charles Scribner, Jr. as quoted in our Section S6 above, with regard to "A Mistranslation of a Passage in Einstein's Original Paper on Relativity," it is imperative that, in the case of this even more important paper by Einstein on his *General Theory of Relativity* (1916), we begin immediately with his original German *Annalen der Physik Band 49* (1916) scientific text, "Die Grundlage der allgemeinen Relativitäts-theorie" (The First Principles of the General Relativity Theory).

This makes it imperative that in the case of the even more general and important original German paper by Einstein we begin immediately with its original German text and its first German page (page 769), with all English translations of it and all later German pages the responsibility of the present writer.

In this task, the first responsibility will be to express our gratitude to the librarian of the Federal Technische Hochschule in Switzerland's Zürich where Albert Einstein achieved his Ph.D. in Physics, for this librarian's xeroxing of the entire 54-page German text of Einstein's most important German treatise, and also for permission to republish here what follows. This German text's first German page (769) opens with the following title and in its first German sentence states the entire German scientific paper's aim and the well-known pure mathematical and mathematical physical aids for achieving it as follows:

17

1916 ANNALEN DER PHYSIK No. 7
VIERTE FOLGE. BAND 49.

1 Die Grundlage
der allgemeinen Relativitätstheorie;
von A. Einstein.[27]

Die im nachfolgenden dargelegte Theorie bildet die denkbar weit-
gehendste Verallgemeinerung der heute allgemein als „Relativitätstheorie" bezeichneten Theorie; die letztere nenne ich im folgenden zur
Unterscheidung von der ersteren „speziellen Relativitätstheorie" und
setze sie als bekannt voraus.

Our English translation is:

The following constructed theory builds on a far-reaching generalization of the present theory usually designated as Relativity Theory;
the latter I call in what follows "Special Relativity Theory" to distinguish it from the former and presuppose it as known.

The remainder of this entire single-paragraphed German page
769 proceeds immediately to tell us with what pure mathematic aids
and with what mathematical physical ones, when applied to the solution of physical problems, he is going to achieve this theoretical mathematical physical aim. They are:

Die Verallgemeinerung der Relativitätstheorie wurde sehr erleichtert
durch die Gestalt, welche der speziellen Relativitätstheorie durch
Minkowski gegeben wurde, welcher Mathematiker zuerst die formale
Gleichwertigkeit der räumlichen Koordinaten und der Zeitkoordinate
klar erkannte und für den Aufbau der Theorie nutzbar machte. Die
für die allgemeine Relativitätstheorie nötigen mathematischen Hilfs-
mittel lagen fertig bereit in dem „absoluten Differentialkalkul,"
welcher auf den Forschungen von Gauss, Riemann und Christoffel
über nichteuklidische Mannigfaltigkeiten ruht und von Ricci und Levi-
Civita in ein System gebracht und bereits auf Probleme der theo-
retischen Physik angewendet wurde.

Our translation:

The generalization of the Relativity Theory becomes very much easier
through the form given it by the mathematician H. Minkowski who
first made clear the formal equivalence of the space-like coordinates
and the time-like coordinate and made this necessary in the construction of the theory. The necessary aids in the construction of the General Theory lay ready at hand in the "absolute differential calculus"
which rests on the investigations of non-Euclidian manifolds by Gauss,
Riemann and Christoffel, and their systematic application to the solution of physical problems by Ricci and Levi-Civita who brought them
into a system.

---

[27] A. Einstein, Annalen der Physik IV, Folge. 49, 1916, pp. 769-822.

Expressed in terms of what Professor Walter Thirring referred to when he wrote in our Section S7 above "Albert Einstein solved the problem of gravitation in 1916," we must expect that on Albert Einstein's next page (770) he will tell us what "the problem of gravitation" was that he "solved in 1916." This he does in two steps. The first consists in telling us what the Classical Theory of Gravitation and his Special Theory (1905) have in common. The second concerns itself with specifying what is unsatisfactory with respect to both the classical and his own 1905 theory. This common unsatisfactory factor defines "the problem of gravitation" to which Professor Thirring referred in the first sentence of his paper without telling us what it was. Hence Einstein's second page of his paper opens as follows:

A. *Prinzipielle Erwägung zum Postulat der Relativität*

1. *Bemerkungen zu der speziellen Relativitätstheorie*

Der speziellen Relativitätstheorie liegt folgendes Postulat zugrunde, welchem auch durch die Galilei-Newtonsche Mechanik Genüge geleistet wird: Wird ein Koordinatensystem K so gewählt, daß in bezug auf dasselbe die physikalischen Gesetze in ihrer einfachsten Form gelten, so gelten *dieselben* Gesetze auch in bezug auf jedes andere Koordinatensystem K', das relativ zu K in gleichförmiger Translationsbewegung begriffen ist. Dieses Postulat nennen wir „spezielles Relativitätsprinzip". Durch das Wort „spezielles" soll angedeutet werden, daß das Prinzip auf den Fall beschränkt ist, daß K' eine *gleichförmige Translationsbewegung* gegen K ausführt, daß sich aber die Gleichwertigkeit von K' und K nicht auf den Fall *ungleichförmiger* Bewegung von K' gegen K erstreckt.[28]

The special Relativity has the following Postulate at its basis which the Galilei-Newtonian Mechanics also satisfies: if a coordinate system K is so chosen that in relation to it the physical laws hold, then *these* laws hold also for any conceivable coordinate system K' that relative to K moves with a constant velocity. This Postulate we call "special Relativity Principle."

By the word "special" is meant that the principle is limited to the case that K' undergoes a uniform translation in motion relative to K, that, however, the equivalence of K' and K does not extend to the case of the non-uniform motion of K' relative to K.

Einstein then turns to the respect in which his Special Theory differs from the classical mechanics of Galilei and Newton (and we add also the classical electromagnetics of Faraday and Maxwell):

---

[28] Albert Einstein *(EGT)* (1916), page 770.

Die spezielle Relativitätstheorie weicht also von der klassischen Mechanik nicht durch das Relativitätspostulat ab, sondern allein durch das Postulat von der Konstanz der Vakuum-Lichtgeschwindigkeit, aus welchem im Verein mit dem speziellen Relativitätsprinzip die Relativität der Gleichzeitigkeit sowie die Lorentztransformation und die mit dieser verknüpften Gesetze über das Verhalten bewegter starrer Körper und Uhren in bekannter Weise folgen.[29]

We translate:

The Special Relativity Theory departs from the classical mechanics not through the Principle of Relativity, but only through the in-a-vacuum Lightquickness Equation Constant, from which, when combined with the Principle of Relativity, the Lorentz Equation for the laws concerning the behavior of rigid rods and standard (Greenwich Time) clocks become relative to the Galilei-Cartesian physical frames of reference in the well-known Special Theory of Relativity.

Then comes the actor's key theatrical cue, which our reading of Professor Thirring's first sentence with understanding has warned us to be looking for:

Die Modifikation, welche die Theorie von Raum und Zeit durch die spezielle Relativitätstheorie erfahren hat, ist zwar eine tiefgehende; aber *ein* wichtiger Punkt blieb unangetastet.[30]

The modification which the theory of space and time undergoes due to the Special Relativity Theory is indeed a deep-going one; but a weighty point remains unsatisfactory.

Quite aptly, it was Ernst Mach, of "Economy of Thought" fame (1885), who suggested to Einstein what this unsatisfactory factor common to all previous modern science and "not less" his own Special Theory (1905) might be, which necessitates the discovery, not achieved by E. Mach, of the new and more General First Principles required:

§2. *Über die Gründe, welche eine Erweiterung des Relativitätspostulates nahelegen*

Der klassischen Mechanik und nicht minder der speziellen Relativitätstheorie haftet ein erkenntnistheoretischer Mangel an, der vielleicht zum ersten Male von E. Mach klar hervorgehoben wurde. Wir erläutern ihn am folgenden Beispiel.[31]

---

[29] *Ibid.*, page 770, 2nd paragraph.
[30] *Ibid.*
[31] *Ibid.*, page 771, Section §2.

§2. *Concerning the First Principles at the Basis of a Generalization of the Relativity Postulate*

The classical mechanics and not less the Special Theory of Relativity contain an epistemological Mangel (defect) which perhaps E. Mach made clearly evident for the first time. We elucidate it in the following example.

Briefly stated, what this elucidation establishes is that the classical mechanics of Galilei and Newton, and "not less" Einstein's own Special Theory of Relativity (1905), are both false, unless both the layman, when he or she looks at natural phenomena, and the observing exact scientist stand on or refer his sensed measured magnitudes of rigid-rod miles and Greenwich-time clocked-seconds to physical K^0 and K^1 objects which are moving relative to one another with constant rectilinear motions. Clearly, since everyone, humanist and exact scientist alike, is usually moving with changing velocities and in non-rectilinear miles this "epistemological defect" (Mangel) pointed out by Mach is a most serious one, affecting not merely every external physical object but each and every human being's way of knowing. Here we have "the problem of gravitation" to which Professor Thirring referred in his classical paper "Gravitation" (1972) but did not describe.[32]

Einstein's solution of this problem is an easy one. *The defect being epistemological, its removal must be epistemological also.* Furthermore, his elucidation noted just above tells him and us what the new Epistemological First Principle must be. It is, as stated on the third page (772) of his original German *Annalen der Physik Band 49* (1916) as follows:

(EEFP-III)   *Die Gesetze der Physik müssen so beschaffen sein, daß sie in bezug auf beliebig bewegte Bezugssysteme gelten.* Wir gelangen also auf diesen Wege zu einer Erweiterung des Relativitätspostulates.[33]

(EEFP-III)   *The Laws of Physics must be so constituted that they hold for any conceivable physical system of moving objects whatever.* This way we go to generalization of the Relativity Postulate.

Einstein then proceeds to show in his next Section §3 *Das Raum-Zeit-Kontinuum* (The Space-Time-Continuum)*: Forderung der allgemeinen Kovarianz für die allgemeinen Naturgesetze ausdrücken Glei-*

---

[32] *Op. cit.*
[33] Albert Einstein *(EGT)*, (1916), page 772, near end of middle paragraph; italics Einstein's.

chungen[34] (The requirement of general covariance for the general tensor of nature's equation for expressing them) that though the Epistemological First Principle (EEFP-III) above is a necessary condition for removing the epistemological defect in all previous theories, it is by no means a sufficient one.

The reason is that both Galilei's and Newton's Laws and Faraday and Maxwell's Laws satisfy Epistemological First Principle (EEFP-III) (being second order differential equations) but do not satisfy Einstein's Special Light Constant c Equation (1905). When the latter prerequisite is satisfied near the end of his Section §3, in his second paragraph on the eighth page (776) of his *Annalen der Physik Band 49*, we have Einstein's second Epistemological First Principle (EEFP-IV).

(EEFP-IV)   *Die allgemeinen Naturgesetze sind durch Gleichungen ausdrück en die für alle Koordinatensysteme gelten, d.h. die beliebigen Substitutionen gegenüber kovariant sind.*[35]

The general laws of Nature must be so constructed through equations which hold for all coordinate systems, that is, that are covariant for any conceivable substitutions (generally covariant).

The speculative discovery and demonstration of these two *Epistemological First Principles,* within the first 8 pages of Einstein's original German *Annalen der Physik Band 49* paper, has one consequence which is so revolutionary that to the present writer's knowledge no one has noted it previously. This is that physics can only remove its own defects by making the science of epistemology more basic than crucially experimentally verified-by-others theoretical mathematical physics: It is precisely in order to emphasize this irreducibility of his two *Epistemological First Principles* to any subject-matter more basic than them that he puts both of them in italics. Moreover, these are the *only* two sentences in his entire *Annalen der Physik Band 49* fifty-four-page paper that he italicizes.

That there is no doubt about this, in Einstein's paper and in his own thinking about exact scientific crucially verified theory in its method. Einstein then uses the remaining 44 (779-822) pages, with their (1) through (4) to (75) rigorously deduced equations to demonstrate. To see why such is the case, we must recall the single para-

---

[34] Because of the Democritean-Eudoxean *First Principle (DEFP-I)* and *(DXFP-II)* we denote this Einsteinean one as *(EEFP-III)*.

[35] *Ibid.,* Section §3, pages 773-778; *Ibid.,* page 776; italics Einstein's.

graphed first German page (769) of his *Annalen der Physik Band 49* paper, where, after telling us what he is going to do, namely, generalize his Special Theory (1905), he then proceeds to tell us with what "pure mathematical aids" (the investigation of non-Euclidean manifolds of the absolute differential calculus by Gauss, Riemann and Christoffel) and with what mathematical physical aids, namely, the method of rigorous proofs for the solution of physical problems of Ricci and Levi-Civita.

Einstein has achieved what he says (in the first sentence of his page 769) he is going to do, when on pages 772 and 776 he has removed the epistemological defect in all previous scientific theory by means of his necessary and sufficient *Epistemological First Principles (EEFP-III)* and *(EEFP-IV)*. These italicized-by-him *First Principles* tell both him and us that in them (to use the language of Sheffer's Definition of Deduction[36]) he, Einstein, has found the irreducible *Minimal Logical Antecedents* of his overall paper's inquiry. It then remains in the remaining 46 (777-822) pages of his paper, using the pure mathematical aids of Gauss, Riemann and Christoffel, and the mathematical physical aids of Ricci and Levi-Civita, to deduce rigorously the equations (1) through (4) to (75) which are the *Maximal Logical Consequents* (i.e., proved theorems) of his two *Epistemological First Principles (EEFP-III)* and *(EEFP-IV)*.

This deduction from these two *First Principles* he begins immediately to demonstrate in the next Section §4 of his *Band 49* (1916) paper. §4 is:

> *§4. Beziehung der vier Koordinaten zu räumlichen und zeitlichen Meßergebnissen. Analytischer Ausdruck für das Gravitationsfeld.*
>
> Es kommt mir in dieser Abhandlung nicht darauf an, die allgemeine Relativitätstheorie als ein möglichst einfaches logisches System mit einem Minimum von Axiomen darzustellen. Sondern es ist mein Hauptziel, diese Theorie so zu entwickeln, daß der Leser die psychologische Natürlichkeit des eingeschlagenen Weges empfindet und daß die zugrunde gelegten Voraussetzungen durch die Erfahrung möglichst gesichert erscheinen. In diesem Sinne sei nun die Voraussetzung eingeführt:
>
> Für unendlich kleine vierdimensionale Gebiete ist die Relativitätstheorie im engeren Sinne bei passender Koordinatenwahl zutreffend.

---

[36] Henry M. Sheffer, *A General Theory of Notational Relativity.* Printed booklet on *Empirical and Relational Logic* (1921). Harvard University Press, Cambridge, Mass.

Our English translation is:

It is not my intention in this treatment to present the General Relativity Theory as a simplest possible logical system with a minimum number of axioms. But it is my main goal to so develop this theory that the reader experiences the psychological naturalness in a persuasive manner and that basic First Principles appear to be learned on the most certain. In this sense, let the following presupposition be introduced:

For an infinitely small four dimensional space, the Relativity Theory in the narrower sense, for any suitable choice of coordinates, holds.

Einstein describes it in German as follows:

Gleichzeitig wird sich die Bewegung des freien Massenpunktes in den neuen Koordinaten als eine krummlinige, nicht gleichförmige, darstellen, wobei dies Bewegungsgesetz unabhängig sein wird von der Natur des bewegten Massenpunktes. Wir werden also diese Bewegung als eine solche unter dem Einfluß eines Gravitationsfeldes deuten. Wir sehen das Auftreten eines Gravitationsfeldes geknüpft an eine raumzeitliche Veränderlichkeit der $g_{\sigma\tau}$. Auch in dem allgemeinen Falle, daß wir nicht in einem endlichen Gebiete bei passender Koordinatenwahl die Gültigkeit der speziellen Relativitätstheorie herbeiführen können, werden wir an der Auffassung festzuhalten haben, daß die $g_{\sigma\tau}$ das Gravitationsfeld beschreiben.

Translated into English this is:

Simultaneously, the movement of a free mass-point is the curved line, such that the Laws of Nature are independent of the moving mass-point. We will also interpret this motion as under the influence of a gravitational field. We note also that the entrance of a gravitational field is joined to a spatio-temporal change of the $g_{\sigma\tau}$. Also in the general cases that we cannot, in a finite region, bring about the validity of the Special Relativity Theory with its usual coordinates, we would hold fast to the interpretation that the $g_{\sigma\tau}$ describes the gravitational field.

Now comes the surprise! The field of gravitational forces is not, as in all previous theories, including even Einstein's Special Equation (1905), identical with the field equations for the Faraday-Maxwell electromagnetic forces. This proved theorem is in Einstein's *Annalen der Physik Band 49*, page 779 at the end of his section §4:

§4 Analytischer Ausdruck für das Gravitationsfeld

Die Gravitation spielt also gemäß der allgemeinen Relativitätstheorie eine Ausnahmerolle gegenüber den übrigen, insbesondere den elektromagnetischen Kräften, indem die das Gravitationsfeld darstellenden 10 Funktionen $g_{\sigma\tau}$ zugleich die metrischen Eigenschaften des vierdimensionalen Meßraumes bestimmen.

§4 The Analytic Equation for the Gravitation Field

Gravitation plays consequently in the General Theory of Relativity an exceptional role as compared with the remaining forces, especially the electromagnetic forces, in that the Gravitation Field's 10 functional field potentials $g_{\sigma\tau}$ at the same time determine the four-dimensional measured metrical properties of space.

Einstein had already noted:

Der Beschleunigungszustand des unendlich kleinen („örtlichen") Koordinatensystems ist hierbei so zu wählen, daß ein Gravitationsfeld nicht auftritt; dies ist für ein unendlich kleines Gebiet möglich. X_1, X_2, X_3 seien die räumlichen Koordinaten; X_4 die zugehörige, in geeignetem Maßstabe gemessene)[1] Zeitkoordinate. Diese Koordinaten haben, wenn ein starres Stäbchen als Einheitsmaßstab gegeben gedacht wird, bei gegebener Orientierung des Koordinatensystems eine unmittelbare physikalische Bedeutung im Sinne der speziellen Relativitätstheorie. Der Ausdruck[37]

$$(1) \qquad ds^2 = -dX_1^2 - dX_2^2 - dX_3^2 + dX_4^2$$

---

1) Die Zeiteinheit ist so zu wählen; daß die Vakuum-Lichtgeschwindigkeit — in dem „lokalen" Koordinatensystem gemessen — gleich 1 wird.

hat dann nach der speziellen Relativitätstheorie einen von der Orientierung des lokalen Koordinatensystems unabhängigen durch Raum—Zeitmessung ermittelbaren Wert.

Our English translation is:

The accelerated state of the "spatial" Coordinate System is thereby so to be chosen that no gravitational field occurs; this is possible for an infinitely small region when X_1, X_2, X_3 become the spatial coordinates, X_4 the co-accompanying physically measured time coordinate (where the unit of time is so chosen, that the Light-in-a-Vacuum Constant in the "local" Coordinate System is put equal to the unit 1). These coordinates have, when a rigid rod is kept in mind as the unit of physical measurement, with a given orientation of the Coordinate System, an immediate physical meaning in the sense of the Special Theory of Relativity. The equation (1) where (1) is:

$$(1) \qquad ds^2 = -dX_1^2 - dX_2^2 - dX_3^2 + dX_4^2$$

In short, from his two *Epistemological First Principles (EEFP-III)* and *(EEFP-IV)*, Einstein has deduced as a proved theorem the c^2t^2 *Light Quickness Equational First Principle* of his Special Theory (1905) for both electromagnetics and gravitation that holds for any

---

[37] Albert Einstein *(EGT)*, *ibid.*, pages 777-778.

Galilean-Cartesian physical frame of reference. We see now why Einstein's *Epistemological First Principle (EEFP-IV)* is necessary.

On the next page (778) of his *Annalen der Physik Band 49* paper he then points out that in the neighborhood of an infinitely small "ortliche" (spatial) magnitude a gravitation field is entailed. Then, by means of two linear homogeneous second order differential equations (proved by "the pure mathematical aids concerning non-Euclidean manifolds" referred to 9 pages back on the first page (769) as the "pure mathematical aids" he will use) Einstein deduced from his two *Epistemological First Principles (EEFP-III)* and *(EEFP-IV)* the following Tensor Equation of Matrices which he numbers (4). His statement is:

> Der Fall der gewöhnlichen Relativitätstheorie geht aus dem hier Betrachteten hervor, falls es, vermöge des besonderen Verhaltens der $g_{\sigma\tau}$ in einem endlichen Gebiete, möglich ist, in diesem das Bezugssystem so zu wählen, daß die $g_{\sigma\tau}$ die konstanten Werte[38]

$$(4) \qquad \begin{array}{cccc} -1 & 0 & 0 & 0 \\ 0 & -1 & 0 & 0 \\ 0 & 0 & -1 & 0 \\ 0 & 0 & 0 & +1 \end{array}$$

The case of the customary Theory of Relativity follows from the considerations it happens to be possible to so choose the physical reference system that it permits the relation between the $g_{\sigma\tau}$ field potentials in a finite region that they take on the constant (i.e., invariant) values of the Tensor of Matrices (4) just above.

Because this equation (4) is deduced by Einstein (with the pure mathematical aids of Gauss, Riemann and Christoffel and the mathematical physical deductive-inductive aids of Ricci and Levi Civita for the solution of the physical problem of gravitation) from his *Epistemological First Principles (EEFP-III)* and *(EEFP-IV)*, we shall henceforth denote equation (4) above as Epistemological Theorem 2. Then, his equation (1) which he deduced first as given at the bottom of our page 25, which is that of his Special Theory (1905), is accordingly denoted as Epistemological Theorem (ET1).

It will enable us to visualize the meaning and import of the latter equation (1) of his Special Theory (1905) in its status as a Special Restricted Case of his General Equation (4) for Gravitational Mechanics, if we make use of the graphical methods of the Einsteinean

---

[38] *Ibid.*, page 779.

26

influenced Sheffer's $n>1$-adic Logic of systems and System Functions which we described in Section S3 above, "The Method of Theoretical Physics."

To this end, by way of contrast, we shall use as an example of such a dyadic relational analytic R_4^2 system the common sense example in which R^2 is the relation of "less than" between the four entitied whole numbers 1, 2, 3 and 4. The result for this System Function is the following Shefferian graph, which we shall denote as (4WN):

	R_4^2	1	2	3	4
	1	− 11	+ 12	+ 13	+ 14
	2	− 21	− 22	+ 23	+ 24
(4WN)	3	− 31	− 32	− 33	+ 34
	4	− 41	− 42	− 43	− 44

Where R^2 = the relation of "less than" between 1, 2, 3, 4 whole numbers.

Let us begin with the diagonal squares. They are 11, 22, 33 and 44. Since no whole number is less than itself, such a relation R^2 is called an all-inreflexive relation R^2. Hence, were we to number these (4WN) R^2 as Einstein has denoted them in his graph (4) above, all four diagonal forms of self-reflexive numbers would be given the value -1; 44 being such, as well as 11, 22, and 33. Such an R^2 relation is called an all-inreflexive relation.

Let us now, following Sheffer's practice in his graduate research seminar in January of 1921, consider the values of the non-diagonal pairs of numbers. They are 12, 13, 14, 23, 24 and 34 to the upper right of the diagonal line in our graph (4WN) above, and 21, 31, 32, 41, 42 and 43 in the non-diagonal pairs and their respective squares in the lower left squares below the diagonal line. Since the relation R^2 of "less than" holds between 12, 13, 14, 23, 24 and 34, but does not hold when the upper left hand squares containing these numbers is imagined to be folded over along the diagonal in our graph above to fall on their converse numbers 21, 31, 41, 32, 42 and 43 on the lower left hand part of the graph above. This entails (since

27

the R^2 relation of "less than" holds for the 12, 13, 14, 23, 24 and 34 numbers but does not hold for their 21, 31, 41, 23, 24 and 34 counterparts) that R_4^2 is what relational analytic R_4^2 logicians call an *all-insymmetrical relation*. Hence, combining their *all-insymmetry* with its *all-inflexiveness*, R in an *all-inflexive all-insymmetrical relation*.

Having thus understood the R where R is the $n>1$-adic (Principle of Relativity satisfying) relation of "less than" between immediately successive whole numbers, the understanding of the $n>1$-adic R where R is the matrix (4) that Einstein deduced in his *Band 49* paper (1916) from his two Epistemological First Principles (EEFP-III) and (EEFP-IV) by the rigorous deduction-induction mathematical physical method of Ricci and Levi-Civita becomes similarly easy. Why? Because only the specific content of the reflexiveness squares and the symmetry squares are different.

These differences become evident if we turn back on Einstein's R equation (4), i.e., to his Epistemological Theorem 2, where Einstein using σσ and ττ pairs of magnitudes where σ denotes spatial and τ temporal magnitudes, each pair of which collectively define the 10 field potentials in his equation (4) of the gravitational field as given and described by him in the matrix as the ττ tensor potential. Filling in the content of these pairs (with what Einstein calls their "constant values"), the following graph for his equation (4), with his deduced epistemological theorem, results:

R_4^2	1	2	3	4
1	$\sigma^1\sigma^1$ −	$\sigma^1\sigma^2$ +	$\sigma^1\sigma^3$ +	0
2	$\sigma^2\sigma^1$ +	$\sigma^2\sigma^2$ −	$\sigma^2\sigma^3$ +	0
3	$\sigma^3\sigma^1$ +	$\sigma^3\sigma^2$ +	$\sigma^3\sigma^3$ −	0
4	0	0	0	τ^4t^4 +

(4)

In this graph the pairs of sigmas outside the principal diagonal line for reflexiveness denote the six symmetrical potentia of the gravitational field. The fact that the six sigma potentials take on + whether they are above or below the diagonal line for reflexiveness expresses the fact that the symmetry of matrix (5) is all symmetrical for its six sigma-metrical field potentials. Because the metrical space of the geometry of Einstein's equation (4) is three-dimensional, Einstein notes immediately a revolutionary consequence. It is that no time-like taus appear in the definition of this metric. This entails that "the Newtonians and the laymen" common sense notion that absolute one-dimensional time flow represents something objectively absolute, within which gravitational forces act, is without any scientific or epistemological warrant ever. The same thing holds for Kant's Categorical a priori form of sensibility absolute one dimensional time flow, since Kant followed Newton and the common sense lay person on this point.

Even more revolutionary are the theoretical mathematical physical and also the epistemological implications of Einstein's deduction of the diagonal line for reflexiveness in our Shefferian graph above. Obvious is the fact that this reflexiveness is of the Medio species, the field potentials for the sigma-sigma portions of the diagonal being inreflexive with the $\tau 44$ potential being positively reflexive. The entire reflexive line with its four field potentials is Einstein's 1905 Theories Equation. In other words it is the Theorem 1 which Einstein deduced from his previously established Epistemological First Principles (EEFP-III) and (EEFP-IV). Even more important this special Theories (1905) Equation can be deduced as his first Theorem 1, only if his second Epistemological First Principle (EEFP-IV) has been established first. Why? The reason is that Newton's 1686 laws of motion satisfy Einstein's First Epistemological Principle (EEFP-III).

Furthermore the positively reflexive $\tau 44$ field potential entails even greater gravitational mechanical and Einsteinean epistemological import. This arises from the fact that in the 1880's the French theoretical mathematical physicist Poisson proved that the $+\tau 44$ field potential, with continuum theory, completely defines Newton's 1686 laws of motion. This convinced Einstein as it does everyone else that an adequate-for-gravitational mechanics must contain ten field potentials rather than merely the single Poisson $\tau 44$ potential of Newton's theory.

Such is the revolutionary impact of Einstein's discovery in his *Annalen der Physik Band 49* (1916), in its first eight pages of his Epistemological First Principles (EEFP-III) and (EEFP-IV) together with his deductions from them, by means of "the pure mathematical aids of Gauss, Riemann and Christoffel and the deductive-inductive mathematical physical aids of Ricci and Levi-Civita to thereby give rigorous proofs as proved Theorems 1 and 2 of his speculatively discovered verified-by-others special (1905) and general (1916) equations. Basic to everything he ever thought, said or did is the irreducible primacy of the epistemological over the theoretical mathematical physical. Hence, it is that this our Section S8's conclusion is our next Section S9's surprise.

S9

Einstein: The World's Greatest Humanistic and Scientific Epistemologist

In a final reflection by Einstein on the import of his entire life work he said:

> The relation between science and epistemology is a remarkable one. Epistemology without contact with science is an empty scheme. Science without epistemology is — if it be thinkable at all — primitive and muddled.[39]

His deductions from his two *Epistemological First Principles (EEFP-III)* and *(EEFP-IV)* gives his warrant for this statement, which, put negatively, is that the contemporary Einsteinean epistemology, because of its "contact with science," is not "an empty scheme." Certainly an epistemology which can deduce 75 equations from its two irreducible *Epistemological First Principles* is far from being "an empty scheme."

The last sentence, however, is more difficult to warrant and to understand. In fact, for anyone who knew the relaxed, modest, open-hearted Albert Einstein over the years through countless conversations, the adjective "unthinkable" and the unqualified "primitive and muddled" seem entirely uncalled for and out of character for this beloved man. All this is, however, a mere "seeming" and as for why it is so, we, not he, must find a warrant. The major warrant becomes evident if we stop reading our own notions of "unthinkable" and "primitive and muddled" and instead define what the difference is between the two sciences he has in mind and about which he has the expertness necessary to warrant his entire last sentence. The one, in Albert Einstein's case, is theoretical mathematical physics; the other, epistemology.

---

[39] *Albert Einstein: Philosopher Scientist* (1948), Volume 7, Paul Arthur Schilpp, editor, *The Library of Living Philosophers, Inc.*, Northwestern University, Evanston, Ill.

The English word "epistemology" derives from the Greek word 'ἐπιστήμη, which means "knowledge". Thus it is that Webster's International Dictionary defines it as "the science which investigates the method and ways of knowing." In short, epistemology is the science of human knowledge *qua* human knowing. Thus it is that any indicative-sentence science is always the complex of two sciences, one of its own subject matter (in Einstein's case theoretical mathematical physics) and the other of epistemology. Furthermore, any Greek word ending in η or α is always feminine gendered. Hence, it is not a mere metaphor but literally the case to say, as all Greek scientists and humanists did, that epistemology, the science of human knowledge *qua* human knowing, is *the Queen of Sciences,* its other equally important factor being *logic,* or λόγος, the Greek gender of which is masculine, the complex of the two coentailing each other.

Consequently, the notion of the one without the other is a contradiction in terms. Hence, Einstein adds that "science without epistemology is — if it is thinkable at all — primitive and muddled."

To see more precisely why such is the case, it is important to note that the word "primitive" has both a complimentary and a derogatory connotation. It was precisely in order to describe its complimentary meaning that we went to such length in our Section S5 above on the natural history stage of scientific inquiry, in which the Aristotelian naive realistic single entity-property *Logic of Predictables* applies, to show that deductive-inductive science is "primitive" and half-baked, not facing the directly sensed data for which it must account unless, after the manner of Aristotle's *Historia Animalium* and *De Partibus Animalium* and Darwin's *Origin of Species* and *Voyage of the Beagle,* it passes through the natural history stage of scientific inquiry before it attempts to achieve either a theoretical statics or a theoretical dynamics of the more mature stages of natural and humanistic epistemological knowledge with their very much more complex Democritean-Eudoxean *Epistemological First Principles* (DEFP-I) and (DEFP-II) as *generalized further* by Einstein in 1916 with his discovery of his *Epistemological First Principles (EEFP-III)* and *(EEFP-IV)* from which, by way of the *Faraday-Maxwell nineteenth-century Field Physics* and the *Galilean* (1636)-*Newtonian* (1686) *Principle of Relativity* (i.e., relational theory of all entities) *particle physics,* the Democritean-Eudoxean *Epistemological First Principles (DEFP-I)* and *(DXFP-II)* derive as *Special Restricted Cases.*

31

Clearly, therefore, to understand the contemporary Einsteinian epistemology, as defined by Einstein's two *Epistemological First Principles (EEFP-III)* and *(EEFP-IV)*, is to understand a scientific *Logic of* 'ἐπιστήμη which is of necessity very complex. Unfortunately, although Einstein describes each one of its disparate pieces in one or another of his many scientific articles and papers, in no one place or treatise does he bring all of these separate pieces together with the "systematic methods of Ricci and Levi-Civita" in any one unitary theory. This makes the two following chapters in the Schilpp-edited *Albert Einstein: Philosopher-Scientist* (Volume 7) imperative reading. These two chapters are Chapter 13, by Victor F. Lenzen, "Einstein's Theory of Knowledge," and Chapter 14, by Filmer S. C. Northrop, "Einstein's Conception of Science." Of them, on page 783 of the above Volume 7 (1978), Einstein writes:[40]

> The essays by Lenzen and Northrop both aim to treat my occasional utterances of epistemological content systematically. From those utterances Lenzen constructs a synoptic total picture, in which what is missing in the utterances is carefully and with delicacy of feeling supplied. Everything said therein appears to me convincing and correct. Northrop uses these utterances as point of departure for a comparative critique of the major epistemological systems. I see in this critique a masterpiece of unbiased thinking and concise discussion, which nowhere permits itself to be diverted from the essential.

Even so, one question regarding this complex Einsteinean epistemology remained up to 1939 without an adequate well-defined relational analytic R answer. It is: What are the formal properties of the relational R between (1) the aforedescribed Anaximanderean boundless all-embracing feeling awarenessed and sense qualitied imageful existential immediate field component of the complex Einsteinean epistemology and (2) its imageless $n>1$-adic R entitied invariant conservation of matter and energy $T_{ik} = T_{ki}$ tensor equational *First Principles* component? This brings us to our next section, S10.

---

[40] F.S.C. Northrop, *"The Importance of Epistemic Correlation in Scientific Method,"* read in September 1939 at the *International Unity of Science Congress in Harvard University* with Professor Philip Franck in the Chair, and Professor Percy W. Bridgman of Operational Definitional fame as Chief Discussant. Later reprinted as *Epistemic Correlations and Operational Definitions* in Chapter IV of F.S.C. Northrop's *The Logic of the Sciences and the Humanities*, Macmillan, N.Y., 1947; reprint 1983, Ox Bow Press, Woodbridge, Conn.

S10

The Importance of Epistemic Correlations in Scientific Method

One of the basic problems in the unification of scientific knowledge is that of clarifying the relation between those concepts which a given science uses in the early natural history stage of its development and those which enter into its final and more theoretical formulation as a verified deductive theory.

Concepts of the former type we shall call, using a term which the present writer takes from Professor C. I. Lewis, *concepts by inspection*, those of the latter type, *concepts by postulation*.

A *concept by inspection* refers to something which is immediately apprehended and gains its meaning from such a factor. *Concepts by postulation* gain their meanings from the postulates or the theorems of the deductive theory in which they occur. Since the concepts in the theorems of a deductive theory are defined in terms of the concepts which appear in the postulates, they also are *concepts by postulation*.

An examination of a specific crucial experiment in physics in conjunction with the deductive theory which it confirms shows this missing factor to be what we shall call "epistemic correlations." The precise manner in which they enter into scientific method must now be indicated by specifying what they co-relate. This becomes evident when we proceed toward the definitions of each.

The important difference between these two types of concepts is that to gain the meaning of *concept by inspection* one designates denotatively something immediately apprehended, whereas to gain the meaning of a *concept by postulation* one examines the postulates of the theoretical mathematical physical theory in question. These postulates in turn being Principle of Relativity $n > 1$-adic relational analytic R, speculatively discovered, and hence *a posteriori*, define not merely their own meaning, but also that of the theorems deduced from them, which in turn, via their epistemic correlations, describe the operational definitions by means of which the theory in question is crucially experimentally verified. Albert Einstein gave expression to this complex way of objectively verified exact scientific knowing in his 1934 paper, "The Method of Theoretical Mathematical Physics," the science of his own expertness, in the following statement:

Pure thought, that is the axiomatic sub-structure of physics, can grasp reality as the ancients dreamed.[41]

The axiomatic basis of theoretical physics cannot be extracted from experience (Erlebnis) (or, we add, by definitional terms of operational definitions only) but must be freely invented.[42]

Upon this surprisingly speculative way of thinking and knowing not merely Einstein but also the quantum physicists Planck, Schrödinger, Bohr, Heisenberg, Pauli and Dirac agree, as our Volume I will show.

This brings us to a crucial question: If the concepts of scientific deductive theory get their meaning through the free play of the imagination of the theoretical scientist and the concepts of both the postulates and their theorems are *concepts by postulation*, then since the data of the theory's subject matter be directly sensed data denoted by *concepts by sensation*, how can such speculative theory be verified?

It is customary, according to the usual answer, to say that one can verify scientific theory referring to scientific objects given by postulation which are not directly observable by deducing from the postulates of the deductive theory in question theorems which refer to what is directly observable. The fallacy in this explanation is that it asks deductive logic to perform the miracle of taking us from postulates stated in terms of concepts of one type to theorems stated in terms of concepts of an entirely different type.

The strong point in the position of the logical positivists is their implicit, although (to the writer's knowledge) not overtly expressed, recognition of this point. The conclusion which they have drawn therefrom is, however, not so meritorious. Unconsciously realizing that the deduction of theorems from postulates, by formal logic or pure mathematics, can never take us from postulates stated in terms of *concepts by postulation* to theorems given in terms by *concepts by inspection* and noting also, quite correctly, that scientific theory cannot be verified without appeal to the immediately observable factors denoted by concepts by inspection — they have concluded that the concepts in the postulates, therefore, must (if formal logic is not to perform the impossible) be *concepts by inspection* also. In other words, the logical positivists' attempt has been made to reduce the

---

[41] Albert Einstein, *The World As I See It* (1934), Covici-Freede, New York, page 32.
[42] *Ibid.* (1934), page 33.

concepts of all scientific objects to nothing but "protocol sentences about sense data" or, with von Neurath and Reichenbach, to the single entity-property naive realistic "physicalism" of the "Ding Sprache."

None of these conclusions follows. Not only is each out of accord with the actual practice of both theoretical mathematical physicists such as Einstein, Planck, Schrödinger, Bohr, Heisenberg, Pauli and Dirac and theoretical experimental physicists such as Lord Rutherford and Arthur Compton, but also their respective *non sequiturs* overlook the fact that what we call "epistemic correlations" are present in all such theory and that it is only because they are present that the experimental scientist knows what experiment to perform or what experimental apparatus to discover and construct in order to put the *concept by postulation* postulates by their proved theorems to *pro* or *con* crucial experimental verification, with but "one negative instance" to the contrary showing the postulates and their theorems to be false tests as Lord Bacon noted vis-à-vis Newton's Four Logical Universal Laws (1686) in the seventeenth century and as Albert Einstein noted in our own century.

Clearly, what such exact science requires is a theory of the exact sciences which will (1) preserve the irreducible difference between *concepts by postulation* and *concepts by inspection;* (2) permit both theorems and postulates to be expressed in terms of *concepts by postulation,* thereby not requiring formal logic to do more than it is capable of doing; and (3) at the same time make possible the verification or testability of deductively formulated theory by appeal to directly apprehensible empirical factors. It is precisely this which the two entity-termed relation of epistemic co-relation provides. Thusly, the logical and other positivists attain the relational analytic R entitled *a posteriori* propositions which the principle of relativity (Theorem 4 above) gives, while also gaining the crucial experimental verification they need, without being forced into their own patently false "protocol sentences about sense data" and other single entity-property "Ding Sprache" reductionisms into which their own previous theories quite unnecessarily have driven them.

Although Professor Henry Margenau's usage, like that of the pragmatic operationalist Professor Ernest Nagel's preference for "rules of correspondence" as the name for these "epistemic correlations," is unfortunate since it suggests that the latter are merely sub-

35

jective constructs by the individual knowing person, nevertheless the friend and colleague of the present writer, Professor Margenau, has added one prodigiously important and indispensable definition to the concept of "epistemic correlations" in the Einsteinean *Epistemological First Principles (EEFP-III)* and *(EEFP-IV)* and their Epistemological Proved Theorems 1, 2, 3 and 4. This contribution is two-fold: (1) these epistemic correlates are not merely two entity-termed relations, but they are also non-one-one two entity-termed relations. This enables one to deduce, because of the non-one-one isomorphism between the *concept by intuition* sense data component of nature and its *concept by postulation* component, the Epistemological Theorem 5 that the sense world suggests but does not contain, either by operational definition or Whiteheadean "extension abstraction," the real world. This subsumes within the Einsteinean *Epistemological First Principle* (EEFP-III) and (EEFP-IV) as special restricted cases the Democritean-Eudoxean *Epistemological First Principles (DEFP-I)* and *(DXFP-II)*.

Margenau's second contribution is equally important. In his classic work *The Nature of Physical Reality*,[43] he demonstrates not merely that these epistemic correlations are non-one-one two-termed relations but that they vary in their content from one limited scientific domain and its theory to another. Also, he specifies for all the diverse logical universal domains what the unique specific content of the diverse epistemic correlations are. This led the theoretical mathematical physicist Herman Weyl to write Professor Margenau in a personal letter as follows:

> You have written not merely a scientifically accurate, but also a beautiful book.

The relation of the complex Einsteinean epistemology to that of ancient Greek exact science brings us to our next section, S11.

---

[43] Henry Margenau, *The Nature of Physical Reality* (1948), McGraw Hill, New York; reprint 1977, Ox Bow Press, Woodbridge, Conn.

S11

Professor Walter Burkert's Classical Paper "Air-Imprints of Eidola: Democritos' Aetiology of Vision"[44]

Presuppose the following *Diels Fragment on Democritos* by the dependable Sextos Empiricos, as given to the present writer in his Yale University Professor Charles Montague Bakewell's *Source Book on Ancient Philosophy*

> By convention (νόμῳ) sweet is sweet, bitter is bitter, by convention hot is hot, by convention cold is cold, by convention color is color. But in reality, there are atoms and the void. That is, the objects of sense are supposed to be real and it is customary to regard them as such, but in truth they are not. Only atoms and the void are real.[45]

Professor Burkert in his classic paper in Volume II of the *Illinois Classical Studies* (1977) proves, by an appeal to independent Greek fragments, that Democritos in the fourth century B.C. discovered what neuropsychological positive and negative feedback experimental cybernetic psychologists today would call an $n = 3$-adic intensional relational *Logic of Psyche,* which is such that, when the Form of the above "atoms and the void" of the external object (o) that "alone are real" acts, via the intervening medium (im), an Eidola Imprint qual*ity* is impressed upon the *Passive Intellect* of the perceiving subject (S^P) such that (S)'s Active Intellect (S^A), by negative and positive feedback, is able to distinguish Eidola Imprint qual*ities*, the Form (a) of which is not that of the "atoms and the void" which alone "are real" from (b) the Form of the Atoms in the void which are identical for all perceivers.

To be sure, all this is merely implicit in Professor Burkert's paper since his method is not that of the logical universal deductive-inductive verified-by-others procedure of the theoretical mathematical physicists today, nor that given in Euclid's *Elements* but instead is that of the imprint by imprint completely inductive historian. This shows in the second paragraph at the top of page 98 of his (1977) *Illinois Classical Studies* paper.

---

[44] Walter Burkert, *Air-Imprints of Eidola: Democritos' Aetiology of Vision* (1977), *Illinois Classical Studies*, Vol. II, Miroslav Marcovich, editor, pages 97-109.

[45] Charles M. Bakewell, *Source Book on Ancient Philosophy* (1907), Charles Scribner's Sons, N.Y.

That such is the case the following statements on page 98 of Professor Burkert's paper begins to make evident.

Democritos, according to Theophrastos, starts from the simple observation, the "appearance in the eye," ἔμφασις: [epistemically correlated in the sense of our Section S11] "in" the pupil of the eye of man or animal, a small picture of the world, and of the observer himself, "appears." This, of course, was generally known, as Theophrastos in another place condescendingly remarks: "As to the appearance in the eye, this is rather a general opinion: for everybody thinks that seeing comes about in this way, by the appearance produced in the eye." In particular, he mentions Anaxagoros for this assumption. This image in the eye seems to be the important link in the process of transmission from the world outside to the seeing individual. For Democritos, this evidently implies two questions: (1) How is this "appearance" produced, and (2) what happens to it after it has entered the eye? In trying to answer this question, he has to rely on his atomic premises, that there is nothing but atoms, different in form or size, moving through the void, hitting each other or getting fixed together in varying arrangements.

Democritos' main effort is devoted to answering the first of these questions. He distinguishes three factors in bringing about the "appearance": there must be (a) a medium between object and eye, (b) some modification of the medium by the object and (c) some means of transport from the object to the eye.

As a medium, "air" is introduced. Air for Democritos, it is a swarm of atoms, not of any particular shape — as in Plato's *Timaeos* — but with a certain limit of size; bigger atoms constitute water, still bigger ones earth. Thus, as air is the finest of all possible media, it is suited best for receiving imprints, as is fine sand contrast to gravel; . . .

. . . Still it is difficult to see how imprints on air could be produced. [Professor Burkert adds with respect to visual imprints] . . . we can hardly blame Democritos for not having invented wave theory.

Professor Burkert then turns to Democritos' first question, "How is this 'appearance' produced?"

With reference to the three (a) "medium between object and eye," (b) some modification of the medium by the object and (c) some means of transport from the object to the eye referred to in the long quotation above, Professor Burkert adds:

The result of this cooperation is a specific process expressed by the verb συστελλόμενον.

After noting that the Luria translation of this Greek verb as "being produced" is the best one, Professor Burkert then adds:

There have been repeated discussions in modern literature of the problem, how the atomistic theories of vision could account for big "imprints" or "images" passing through the pupil; Democritos has answered it by the concept of συστέλλεσθαι.

Then in the final paragraph on his 4th page (100), Professor Burkert turns to Democritos' second question, which is:

> But how, finally, are the imprints transported? This too is expressly stated in the text of Theophrastos. This makes perfect sense. We know that Democritos dealt with problems of perspective, and gave some explanations of how we can "see" magnitudes and distances correctly, though Theophrastos did not deign to describe it (54, p. 514 f. Dox.). He must have assumed that the size of the imprint is proportionally reduced according to the distance from the eye. In this process of "shrinking," "that which sees" plays some role. Democritos apparently supposed that, as in the case of the owl's eye, though with less force, fire-atoms constantly emanate from the eye, like the "cone of visual rays" of later optics. Later on, the Stoics speak of a state of "tension," συνέντασις, of the air produced by the cone of visual rays. The fire-atoms of the eye somehow cut a path through the air along which, then, the "imprint" is transported, shrinking all the time in the cone.

With respect not merely to the $n = 3$-adic (o) (im) (S^P and S^A) cooperation between (1) the (νόμῳ) nominalistic radical empirical Democritean γῆ-earthy τὸ ἄπειρον undifferentiated field continuum factor and (2) its epistemically correlated $n = 3$-adic (o) (im) (S^P–S^A) relational analytic R actional *Logic of Psyche* factor, Professor Burkert notes on page 98 of his classical paper that:

> Democritos has to presuppose an interplay of three activities: an "efflux" (ἀπορροή) of the object, some action of the eye and, an impulse coming from the sun or light. This third factor is easiest to explain: fire-atoms, being smallest and swiftest, are constantly emitted by every source of light; . . . This third factor, some activity of "that which sees," has sometimes seemed to be suspiciously close to Plato's theory of the active eye;[46]

With these inductive historical facts established, Professor Burkert then reaches the climatic conclusion of his paper:

> Finally, there is the famous theory of *eidola*, which has its place in every handbook of the theory of vision of atomism ever since Lucretius. "External objects are constantly throwing off films of atoms which retain the approximate form of their surfaces and so constitute 'images' of them. These actually enter the eye. . ." Doxology ascribes this theory

---

[46] Burkert (1977), *op. cit.*, page 99.

to "Leucippos and Democritos"; we have the title of a book of Democritus, Περὶ εἰδώλων ἢ περὶ προνοίης (68 B 10a = 578 Luria) and of a polemical work of Heracleides Ponticus, Περὶ εἰδώλων προς Δημόκριτον. In fact, Theophrastus, in the main text discussed above, refers to this theory: "On the whole, if one assumes an efflux from the form, as in the book Περι τῶν εἰδῶν, why should one bring in the imprinting process? For the images, by themselves, appear in the eye." It is generally agreed that Theophrastus, in the parenthesis, is referring to some special book of Democritus, [As noted] . . . Democritus has to presuppose an interplay of three activities: an "efflux" (ἀπορροή) of the object, some action of the eye, and an impulse coming from the sun or any other light. This third factor is easiest to explain: fire-atoms being smallest and swiftest, are constantly emitted by every source of light; hitting the air, then produce a "condensation," as children playing in sand "condense" it by heaping and pressing it; if wax is too soft it will not keep any imprint.

In this connection, Professor Burkert also notes that: "Understandably, Democritos has much less to say on the second main question, what happens to the 'appearance' when it has come to be in the eye. Evidently it does not stay there, but is transmitted to the 'rest of the body.' Soul, for Democritos, is not concentrated in any 'leading' organ." Instead, this complex changing "soul," Professor Burkert adds, is the effect of the interplay of the external object (o), via the intervening medium im, and the passive intellect S^P active intellect S^A, to arrive at a neuro-cybernetic trial and error theory of the external object (o) that is known. All this enables Democritos to escape the single-entity property disembodied Ghosts of the modern idealists on the one hand and the feelingless and thoughtless material substances (Ghasts) of the modern materialists on the other hand.

Positively stated, when combined with the defined theory of books VII and VIII of Euclid's *Elements*, which were discovered and defined by Democritos and Archytas together with the latter's friend Plato, slightly later, discovered also an epistemological First Principle which they described as the thesis "the sense world suggests but does not contain the real world." This principle we denote (as above, p. 31) as (DEFP-I). With the discovery of arithmetically incommensurable geometric magnitudes, the epistemological giant Eudoxos, in the Definition 5 of Euclid's Book V, in its concept of *Logos* achieved the more general epistemological principle from which the previous Democritean one follows as a special restricted case. This more general Logos Democritean-Eudoxean theory we shall denote as (DXFP-II).

Although in this "similarity," Professor Burkert does not refer to the Greek *'EN APXAI*-defined Democritean, Archytean and Platonic Books VII and VIII in Euclid's Greek *Elements,*[47] to do so is to realize this "similarity" is no mystery. Put in this wider Greek Euclidean historical and $n = 3$-adic's shared *'EN APXH* acoustical and astronomical verified common scientific context, as given us in the Definition *20* of ἀναλόγίαι, *proportions,* meaning as explained by Heath, "the arithmetic, geometric, and harmonic, where of course the reference is to arithmetic, geometric and harmonic *means* (μεσότητες)," which elemental $n = 3$-adic *R 'EN ARXH*, Democritos (c. 469 B.C.–370 B.C.), as given in Euclid's Book VII, is the discoverer, constructor, and, in the science of acoustics (Greek music), the verifier, with Plato's most intimate friend and scientific colleague Archytas (fl. 380 B.C.).

Combining these two Greek *Epistemological First Principles* with the two more general *Epistemological First Principles* established by Einstein in the first eight pages of his *Annalen der Physik Band 49* paper (1916), and then adding another discovered and published by Heisenberg in 1948, the warrant for which must await our analysis of the indeterminacy quantum mechanical theory in Part B of our forthcoming Volume I, we arrive at the following six species of Epistemological First Principles, each succeeding one subsuming within itself its immediate predecessors:

(NHSI)   The *Aristotelian* naive realistic *Genus-Species Logic of Predictables.*

(DEFP-I)   For any two sensed measured magnitudes whatever, there are always two n>1-adic relational analytic *R* smaller magnitudes which are not sensed, and there is always a smallest atom's diameter magnitude, which is not sensed. (This is the First Postulate of Particle Physics.)

(DXFP-II)   For any two measured magnitudes which are sensed, there are always two $n>1$-adic relational analytic *R* magnitudes which are not sensed, and there is never a smallest magnitude. (This is the basic epistemological First Principle of Field Physics.)

(EEFP-III)   The Laws of Physics must be expressed by equations which hold for all coordinate systems, i.e., remain covariant (generally covariant) for any possible transformation of coordinates.

---

[47] T.L. Heath, *The Thirteen Books of Euclid's Elements* (1908), *op. cit. H-II,* pages 278-344 and 345-383.

(*EEFP-IV*) The Laws of Physics must be so constructed that they must hold for any moving physical system of entities whatever.

(*HEFP-V*) *The Epistemological First Principle of Limited Logical Universal Domains*. The import of (HEFP-V) has been completely overlooked; otherwise a contradiction exists between (1) the 100% deterministic mechanical causality of Einstein's General Equation (EGT) (1916) and (2) the Qualified Mechanical Causality of the Indeterminacy Principle Quantum Mechanical Equation.

Consider the following question. Could it be that the reason why Albert Einstein, in the two Brussels Solvay conferences in 1927, and even on to the end of his life in 1955, could not rejoice with his beloved quantum theoretical colleagues over the verification of their equation as they rejoiced over the verification of his in 1916 is that he never became aware of a paper published by Heisenberg, upon Pauli's insistence, in a journal named *Dialectica* in Neuchatel, Switzerland in 1948? It would seem so.

In any event, Werner Heisenberg's *Principle of Limited Logical Universal Domains* removes the contradiction which divided these great giants. For Einstein's 1916 equation's *domain* is that of *gravitational mechanics,* and the quantum theory's *domain* is that of *electromagnetic dynamics* and of *interchemicalatomic interactions.*

Even with the above six different epistemological species of *Epistemological First Principles Science* kept constantly in mind, as we must do in everything which follows, we shall face a quite different ordinary linguistic problem which must be overcome. So complex is the difficulty that it requires a section unto itself.

S12

Three Ordinary Language Difficulties in Our Way and Their Removal

Alfred North Whitehead made me aware of this difficulty in September of 1922, with the following warning, as we began the investigation of the identity and differences between his epistemology and that of Albert Einstein:

One cannot be too suspicious of ordinary language in science and philosophy, the reason being its single entity-property syntax.

Whitehead met it with his "knowledge by adjectives-knowledge by relatedness Method of Extension Abstraction" epistemology.

Our question now becomes: Can the Democritean-Eudoxean-Einsteinean-Heisenbergean epistemology meet it also? To this question we must now turn. The difficulties involved become most evident when one notes that all these theories use, and of necessity must use, ordinary language with its single entity-property syntax in the statement of what the symbols in their elementary equations mean. For the humanities, this ordinary language syntax is even more difficult to escape. In the case of Einstein's general *Annalen der Physik Band 49* scientific paper, his two italicized *First Principles* (from which his Equations (1) through (75) are deduced) are stated in the ordinary German language.

It took the present writer from 1922 to the publication of a paper in 1970 (the J. E. Smith edited Chapter 5 of *Contemporary American Philosophy,* second series [1970)])[48] to describe the complication of this epistemological predicament and specify how to overcome them. This publication opens thus:

> In *The Two Cultures,* C. P. Snow describes our present predicament with respect to this topic. It is not a very happy one. At best, scientists and humanists find it difficult to communicate with one another. At worst, bitterness occurs. Witness the inhumane *ad hominum* attack upon Professor Snow by one of his humanistic colleagues at his University of Cambridge. It has not always been so. It need not be so today. To suggest why for both the past and the immediate future is the purpose of this investigation.
>
> The inability to communicate provides a sufficient reason for our unfortunate situation: neither scientists nor humanists specify the more elementary common denominator factors which are necessary to relate the specific contents of successively generalized theories of naturalistic scientific knowledge in Western culture to the *specific contents* of its various humanistic intrinsic values. Were this not so, bitterness at least could be avoided. And, although the reduction to common denominator factors and its communication present their difficulties due to (a) the various intrinsic values (aesthetic, moral, officially legal, and religious), (b) the successive increasingly generalized scientific theories and especially (c) the ambiguities of ordinary language, there need be no insurmountable *difficulty* in the way *in principle.*

By "intrinsic values" in the sentence just above, we do not mean the single entity-property categorical or hypothetic *a priori* Ghostly one of Kant, Hegel, the Hegelean Marxist or the Popperean Neo Kantian. Nor do we mean with our tutor G. E. Moore that the word "good" is indefinable; even though he in his Scottish realism affirmed

---

[48] F.S.C. Northrop (1970), *Contemporary American Philosophy,* second series, J.E. Smith, editor, page 107.

that Hume and Kant were right when they showed that no *is*, be it naturalistic or humanistic, can give an *ought*. Unfortunately, however, from this valid affirmation, all of the above moralists inferred a *non sequitur*; the *non sequitur*, which the present writer pointed out in his *Delaware Seminar Lecture* (1961),[49] namely, that though an *is* can never be a sufficient condition for making an obligatory judgment, it by no means follows that it is not a necessary condition — the necessary condition moreover that must be satisfied before the evaluation judgment can even begin.

The science of Jurisprudence makes this unequivocally clear. The habit of moralists of overlooking "the facts of the case" minor premise in the legal judgment that requires the judge's "charge to the jury," in the jurisprudential syllogism, and acting as if the major premise is all that exists in human interpersonal moral, legal or religious evaluations is the universal error and practice of today's demoralized humanistic world. The removal of this *non sequitur* is the reason why this Prolegomena must be followed, after we have specified the verified First Principles of the *is* in today's remarkable verified theoretical mathematical physical and neuropsychological theorem in our forthcoming Volume I, with Volume III entitled *et Jurisprudencia*.

Hence, also, here and now we must face this Section S12 further.

Our first difficulty is that we have to use Aryan ordinary language to correct the countless errors its (*a*) two-termed entity-property syntax and (*b*) diverse confused and other meanings (picked up over the centuries by each one of its words) have produced. As Whitehead emphasized to me in 1922 and repeated annually afterwards: 'One cannot be too suspicious of ordinary language in science or philosophy.' Although it is a rich vehicle of expression, it is an exceedingly slippery and treacherous one.

In value subjects, the fatal dangers are much greater than in natural knowledge. For as noted above, Newton falls into grave error in his ordinary language writing, only to be saved by the different many-entity-termed relational syntax of his mathematical language. But most medieval and modern Western humanists at least use only one or another of the Aryan ordinary languages. Also, even though some lawyers have avoided the pitfalls of Aryan prose better, as the sequel will show, than other social scientists and humanists, these lawyers do not use the imageless relational symbolic language of mathematics. Consequently, if Einstein found physics 'primitive and muddled' due to the neglect of epistemology, we must not be surprised to find the humanities and 'social science' triply and quadruply so.[50]

---

[49] F.S.C. Northrop, The *Delaware Seminar* (1961), William L. Reese, editor, page 1-19.
[50] *Ibid.*, pages 6-19.

How are we to meet and overcome this difficulty? The answer has three parts. First, because all sensed or introspected qualities are disjunctively related rather than necessarily connected, as Hume showed, the logic of the (i) *differentiated* nominalistic radical empirical component of cognitive knowledge is the extensional propositional calculus of the Stoics[51] and *Principia Mathematica*[52] with its 'material implications,' and the later Wittgensteinean Truth-Tables (1923).[53]

This the remarkable Charles Saunders Peirce called *Secondness*. It entails the ordinary language rule which we shall call:

Maxim I: When referring to the (i) component of cognitive knowledge, all its words must be thought of, written, and read *only* in their nominalistic radical empirical meaning, as described best for the so-called 'outer' sensed qualities by Berkeley, and for introspected qualities by the skeptical Hume of Volume I of *The Treatise* and in William James's *Radical Empiricism* and other essays on the self.

Our second difficulty (ii) arises from the (iii) many-entity-termed logical realistic factor in warrantable knowledge of human and cosmic nature. Overcoming it gives ordinary language *Maxim II: When referring to (iii), the noun in our ordinary sentence must be an irreducible many-entity-termed relation and its predicate the substantive content of that relation.* This Maxim has some surprising consequences.

Corollary I: Applied to the two-termed entity-property $\phi_x \psi_x$ symbols of *Principia Mathematica*, it entails that the ϕ_s and ψ_s must always be many-entity-termed relationally defined. This avoids the 'set-theoretical' paradoxes; without either the *ad hoc* restricted theory of types of Earl Russell and Professor W. V. O. Quine or the weak kind of self-reference of the unique *Symbolic Logic* of Professor F. B. Fitch, which, when its propositions add the additional postulate necessary for specific content, loses sufficient self-reference and then falls back into the *ad hoc* ness of the theory of types. The unqualified validity of Godel's theorem needs also to be re-examined. Legal science will give reasons for this also.

Corollary II: Necessary, too, is the construction (only now in process) of an irreducible at least three-entity-termed logic of relations. Its intentional syntax is that of: The Stoics *Laktoa*, Descartes's self-referential *Cogito*, Kant's regulative moral ego, Brentano and Husserl's act psychology, von Domarus's embodied psychiatry, and Drs. (M.D.) Warren

---

[51] B. Mapes. *Stoic Logic* (1955). University of California Press, Berkeley, U.S.A.

[52] Whitehead and Russell, *Principia Mathematica* (1905). "Theory of Deduction", Cambridge at The University Press, England.

[53] Ludwig Wittgenstein, *Tractatus Logico-Philosophicus* (1923). London, *ibid.*, *Tractatus Logico-Philosophicus* PB (1921), Routledge and Kegan Paul Ltd. London. Translated from the German by C.K. Ogden with an Introduction by Bertrand Russell.

S. McCulloch, Arturo Rosenbleuth and Valentino Braitenberg's experimental cybernetic (i.e., self-referentially *helmsman*ing), circular ('feedback') mechanically causal neurophysiological systems, as well as the irreducible three-termed relations of Royce's 'Principles of Logic' and Peirce's *Thirdness* (CPCP). Until such a constructed intentional three-termed logic is achieved, ordinary language will serve as a makeshift providing we realize that it is 'unthinkable without epistemology' and follows *Maxim II* above.

Reference was made above to 'the (i) *differentiated* nominalistic radical empirical component' of human nature and natural knowledge. The word *differentiated* brings us to our third linguistic difficulty.

Both James and Peirce (and also the ancient Oriental nihilistic Mahayana Buddhists and the unqualified non-dualistic Vedantic Hindus) noted that the totality of nominalistic all-embracing existentially felt radical empirical immediacy is not everywhere and everywhen differentiated into successive 'perpetually perishing' qualities. Instead (for the radical empirical knower and evaluator as for the exact theoretical and experimental natural scientist), all of these Buddhist unqualified non-dualistic Asians (both South and Far Eastern) as well as James and Peirce noted, to use James language, that it is only at "the focus of attention" that (νομῳ) nominalistic γῆ-earthy τὸ ἄπειρον (boundless) subject (*s*) intervening medium (*im*) external object (*o*) presentational immediacy is sharply differentiated and qual*itied* and that, positively given, is "at its periphery" vague, indeterminate and undifferentiated, a *per se* bare *quale* — which Peirce called *Firstness,* and the present writer a half-century later (1946) in his *The Meeting of East and West* called the "undifferentiated aesthetic continuum." We present Peirce's pointer-word Firstness first.

> . . . The idea of the absolutely first must be entirely separated from all conception of or reference to anything else; for what involves a second is itself a second to that second. The first must therefore be present and immediate, so as not to be second to a representation. It must be fresh and new, for if old it is second to its former state. It must be initiative, original, spontaneous, and free; otherwise it is second to a determining cause. It is also something vivid and conscious; so only it avoids being the object of some sensation. It precedes all synthesis and all differentiation; it has no unity and no parts. It cannot be articulately thought: assert it, and it has already lost its characteristic innocence; for assertion always implies a denial of something else. Stop to think of it, and it has flown! What the world was to Adam on the day he opened his eyes to it, before he had drawn any distinctions, or had become conscious of his own existence — that is first, present, immediate, fresh,

new, initiative, original, spontaneous, free, vivid, conscious, and eva-
nescent. Only remember that every description of it must be false to
it.[54]

To this we urgently add: Remember also, that "*it*" is not an *it*.
So much for Peirce's pointer-word *Firstness*. The indeterminacy
principle wave mechanical quantum physicist, who possesses also a
remarkable intuitive comprehension of the Buddha's Nibbana and
the Hindu's unqualified non-dualistic Vedanta's "Psychic Atman
which is the cosmic field continuum Brahman" of all Buddhist South
and East Asian, and especially Zen Buddhist cultures, denotes
Peirce's *Firstness* (as attempted above) thusly in his (Schrödinger's)
What Is Life?,[55] where after stating that "the pluralization of con-
sciousness, or minds . . . leads almost immediately to the invention of
souls as many as there are bodies," then adds that "the only possible
alternative is simply to keep to the immediate experience that con-
sciousness is a (field continuum) singular."

We add that, being as Peirce denotes "indescribable" (apart from
its differentiated somewhens and somewheres), its logic must be that
of the above Asians' *Neti-Neti* (it isn't this differentiated quali*ty*, it isn't
that differentiated quali*ty*.) Otherwise, neither (as John Dewey also
saw) can radical empirical feeling-awarenessed man or woman nor
even nature's impressionistic beauty be kept within natural knowledge.

So much for Peirce's *Firstness* pointer-word. Because our alternative pointer-word meets a difficulty Peirce's *Firstness*, *Secondness* and
Thirdness never faces, he left the relation between the three factors
(a)(b)(c) "in our way" undefined. It is to be recalled that Whitehead's
"Method of Extension Abstraction" had no such weakness. It
specified the "knowledge by adjective" and "by relatedness" by which
Firstness, *Secondness* and *Thirdness* relate to one another in verified
indicative sentence human knowledge.

The same thing needs to be done for the much more complex
Democritean-Eudoxean-Einsteinean epistemology. This was done in
1946 by the present writer in his *The Meeting of East and West*. The key
to success centers in the fact that the (a)(b)(c) factors are related to

---

[54] *Collected Works of Charles Saunders Peirce*, edited by Charles Hartshorne and Paul Weiss, Harvard University Press, Cambridge, Mass.

[55] Erwin Schroedinger, *What Is Life?* (1946), Cambridge University Press, pages 94-95.

one another by an *epistemic correlation* (in the sense rigorously defined in this *Prolegomena's* Section S10 above) in the precise manner in which Figure II below (from page 453 in *The Meeting of East and West*) is to be distinguished from the single entity-property syntactical (1) material substances → (2) material substance → (3) phantasmic projections erroneous notion with all its insolvable non-existent "body-mind" and other problems, and also the "set theoretical" and "Hempel paradoxes" entailed by Peirce's *Secondness*.

To understand why Peirce escapes these erroneous notions, rendering all traditional modern philosophical theories obsolete (the nothing-buttery positivism, the disembodied "Idealists'" Ghosts and the Materialists' feelingless and thoughtless Ghasts alike), we need to recall another statement made by Alfred North Whitehead to the present writer in the fall of 1922. It was:

> A mistake was made in the interpretation of the entities of modern physics at the beginning of the modern world, and it is only by returning to the origins and correcting this error that any solution for the problem of traditional modern philosophy in science or the humanities is to be found.

Figure I specifies the error; Figure II removes it.[56]

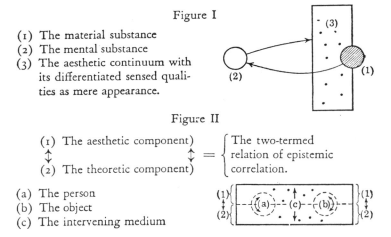

Figure I

(1) The material substance
(2) The mental substance
(3) The aesthetic continuum with its differentiated sensed qualities as mere appearance.

Figure II

(1) The aesthetic component) ⇕   = { The two-termed relation of epistemic correlation.
(2) The theoretic component) ⇕

(a) The person
(b) The object
(c) The intervening medium

Because furthermore, as has been shown in our Section S2 on Einstein's statement concerning the indispensable import of Euclid's Geometry, if today's physics is to be understood (a thesis in which Whitehead also concurred) and especially because of Professor

---

[56] F.S.C. Northrop, *The Meeting of East and West, op. cit.*, page 453.

Burkert's paper on Democritos, the following topic takes on considerable importance, so much so that it demands a section unto itself.

S13
Two Independent Publications in 1970

The first is by the present writer and sets the stage for the other more important achievement. Hence, we shall describe briefly the former publication first. It has to do with *Corollary II* of our *Linguistic Maxim II* which, let it be recalled, is that: *the noun in one's ordinary language indicative sentence must always be an* n>1-*adic relational analytic* R *and the indicative sentence's predicate must always be the formal defining properties of that specific* R.

This entails that whenever one finds, in the ancient Euclidean verified Elements or any other $n>1$-entity-adic ancient Greek Democritean-Eudoxean-Euclidean Greek verified natural or humanistic science, the two Greek words *EN APXAI* or *'EN APXH*, or Εν αρχη, or the word "In Principle" in today's Einsteinian-Heisenbergian epistemological verified science, these two words are always followed by the Greek definite article ὁ, meaning "the one, and the only one which," such that what follows the single lettered word ὁ gives the $n>1$-adic defining properties of the *'EN APXH* in question.

This entails, as the ancient Greeks and Einstein and Heisenberg knew very well, that, being thus relational and analytic *R* entitiedly defined, there is in all such Western science or the humanities no synthetic categorical or hypothetical *a priori* knowledge; all verifiable human knowing, whether in natural science or the humanities (i.e., aesthetics, jurisprudence or theology), being *a posteriori*. Such is the epistemological reason why the present writer in his (1970) essay's *Corollary II* of *Linguistic Maxim II* wrote as follows:

> *Corollary II:* Necessary too is that of: the Stoics' *Lekton*, Descartes' self-referential *Cogito*, Kant's regulative moral ego, Brentano and Husserl's act psychology, the research (M.D.) of Dr. Eilhard von Domarus' Yale Ph.D. thesis in Philosophy (1932) on *The Logical Structure of Mind*, published later in *Communication: Theory and Research*, ed. Lee Thayer, and Drs. (M.D.) Warren S. McCulloch, Arturo Rosenblueth and Valentino Braitenberg's experimental cybernetic (i.e., self-referentially helmsmaning) (due to Norbert Wiener) circular mechanical causal negative

and positive ('feedback') neuropsychological systems, as well as the irreducible three termed relations of Royce's *Principles of Logic* and Peirce's *Thirdness*. Until such a constructed intentional three-termed logic is achieved, ordinary language will serve as a makeshift, providing we realize that it is 'unthinkable without epistemology,' and follow our Linguistic Maxim II above.

Obviously, the full exposition of all the foregoing factors and their natural history clinical comparative neuropsychological data must await the later occasion of Book XIII in this our present *Prolegomena*'s forthcoming Volume I. It suffices to note here two factors: the first one is that all the persons, ancient Greek, traditional modern and present-day, referred to in the above condensed *Corollary IId* are involved. Second, the greatest advance was made by the group, all of whom were in the World War II Macy Foundation for Medical Research conferences chaired by the Drs. (M.D.) Warren S. McCulloch and Frank Fremont Smith, of which the present writer was a founding member. The greatest advance of all was made by the then young natural history clinical research M.D. from the University of Freiburg in West Germany's Schwarzwald, Dr. Eilhard von Domarus, as given in his Yale University Ph.D. thesis, *The Logical Structure of Mind*. With respect to this remarkable achievement, it is sufficient to note here and now that von Domarus, under the direction of the present writer in 1923, investigated the identities and the differences between Whitehead's equations and their less speculative natural history based data epistemology because they, being less speculative, were closer to his (M.D.) University of Freiburg comparative clinical data, instead of following the different equations and the more speculative epistemology of Einstein.

Thus, in the Preface to this epoch-making thesis *The Logical Structure of Mind* (1929), as rewritten in 1967 after further verification with Dr. Warren S. McCulloch in New York State's Rockland State Hospital, von Domarus tells us two things. First:

> The way in which the question (of the logical structure of mind) arose suggested that the solution was to be sought in the nature of experience itself. I was at the time in Freiburg and turned to Husserl in phenomenology. My personal indebtedness to him is greater than appears in this thesis. He was concerned with mentality and clarified for me the notion of transcendency in terms of intention. On account of his teachings, my study in psychology began, though it did not end, with the writings of Brentano. Hence, the chapter on "The Sciences of Mind." [58]

---

[58] Eilhard von Domarus, *Communication Theory and Research* (1967), Lee Thayer, editor, Charles C Thomas, Springfield, Ill., pages 351-411.

By an unusual circumstance in this very year (summer of 1922), unknown to one another the present writer was in Husserl's seminar and also his first lecture, which Husserl opened with the words: "Descartes is the father and Brentano is the grandfather of the 'reine Beschreibung' method of phenomenology." Also Husserl had his then private docent Metzger take the present writer, sentence by German sentence — each sentence "sehr wichtig" of the first eleven pages — the "Wichtigkeit" taking Herr Dr. Metzger so long to explain each one of Husserl's *Ideen*. Clearly, these "reine Beschreibung Ideen" were not the nominalistic ideas of Hume, but the *Logical Universal First Principles Ideen* of Brentano's Telelogica and Descartes' continental rationalistic *self-referential* intentional logical *Cogito, Ergo Sum*.

Von Domarus' second statement in his 1924 Ph.D. thesis was:

> But the bridge, so defined, was not substantial. What was requisite was a theory of knowing. I came to Northrop for the philosophy of science. His chief interest was in understanding the known in terms of how the known arose and of how knowledge of it was systematized in terms of logical relations. Through him, I became aware of Whitehead and the development of mathematical physics. While studying with Northrop, the problems of understanding, as of ontological importance, particularly on biology, became clear to me. At his suggestion, I wrote this thesis as a dissertation for the Ph.D. in the Department of Philosophy in Yale University.

It suffices to note that because its less speculative Whiteheadean "Method of Extensive Abstraction" was closer to the natural history comparative psychological data, von Domarus based his logical structure of mind on the Whitehead equations and epistemology rather than on those of Einstein.

This was a wise decision, for it took neurophysiological psychologists some 43 years between the von Domarus Ph.D. thesis and 1970, when an $n = 3$-adic intensional relational logic of neurophysiologically embodied psyche, based on the much more speculatively complex verified-by-himself and others Einsteinean-Heisenbergean (EEFP-III), (EEFP-IV) and (HEFP-V) epistemology, was discovered. This brings us to the second independent publication in 1970.

In the light of (1) von Domarus' achievement of the late 1920's and (2) the Macy Foundation for Medical Research Cybernetic Feedback Conference in the early 1940's, which was chaired by Warren S. McCulloch M.D. and Frank Fremont Smith M.D., in which Norbert Wiener, John von Neumann, Arturo Rosenbleuth, Julian Bigelow and the present writer were founding members, we need not be surprised that the author of this second publication in 1970 brought to his achievement three skills. First, he had an M.D. from the same natural history medical school in comparative clinical psychiatry in the University of Freiburg in which von Domarus was trained some 20 years earlier. Second, he spent a year associated with Warren S. McCulloch doing research on experimental cybernetic neurophysiology in the Massachusetts Institute of Technology in Cambridge, Massachusetts, and third, he is the Ph.D. director of the Neurophysiological Laboratory of the medical school of the Universidade de Lisboa, Portugal, while also being Professor of Psicologia Medicina.

The author of this complex achievement is Professor José Simões da Fonseca M.D., Ph.D. The title of his (1970) publication is *Signification and Intention*, J. Simões da Fonseca, editor. The titles of this 97-page volume's Chapters I through VII are: *I — Psychological Processes from the Viewpoint of Signification and Intention, II — A Calculus of Signification and Intention* (this is what everyone had been seeking: De Morgan (1863), Peirce (1900), Royce (1905), McDougal (1912), Husserl (1916), Craik (1943) and McCulloch (1943)), *III — Neural Nets from the Viewpoint of Signification and Intention, IV — Learning as a Change of Signification and Intention, V — Cognitive Processes: Some Means of Formal Representations, VI — Language from the Viewpoint of Signification*, and *VII — Psychiatry from the Viewpoint of Signification and Intention.*

Thanks to our Section S11 on Democritos' $n = 3$-adic relational analytic R entitled $(o){\mapsto}(im){\mapsto}(S^p) - (S^A){\mapsto}(im){\mapsto}(o)$ τὸ ἄπειρον γῆ-earthy Logic of Psyche, we can understand Professor Fonseca's word "signification" and its role in human knowing. Given are the Eidola Imprints in the passive perceiver's (S^P) eye; "signification" then is what the particular perceiver's (S^A) active intellect (positive and negative feedback-wise) interprets the form of the passive intellect's Eidola Imprint as signifying external object (o)-wise. Intention then, as used by Professor Fonseca, refers to the Signified Form of the Object as defining the intent or purpose of the perceiver's motor neuronic immediate or subsequent behavior.

The heart of the matter centers therefore in having discovered an $n = 3$-adic intensional relational logical calculus which can account for how the perceiving subject (S), given the form of the *Eidola Imprint,* can, trial and error (negative and positive) feedback-wise, infer, via the intervening medium (*im*), the veridically verified relatedness of the objects known. It is the great merit of the Trinity Cambridge psychologist, the late Kenneth Craik,[59] to have foreseen that only $n = 3$-adic relational theory of knowledge can do this. Clearly this, the conception of the objects (o) as single entity-property feelingless and thoughtless Ghasts, and of the perceiver as a pluralistic or monist Ghost, cannot possibly provide. Thus it is that the Fonseca Calculus of Signification and Intention *ipse facto* dismisses as impotent both the three Freudian disembodied Ghosts in the basement of everyone's *"death-wish" psyche* on the one hand and the feelingless and thoughtless single entity-property Ghosts of the Skinnerite behaviorists on the other hand. Stated positively, therefore, what the Fonseca Calculus of Signification and Intention (and it must do both, for Peirce and McCulloch failed probably because they got a theory of what the Eidola Imprint *signifies only*) does is to specify an $n = 3$-adic *Intensional Relational Logic of Neural Net Embodied Psyche,* where its three components he denotes as (x), (B) and (y), such that (x) is an intensional logic of relations of the Sheffer-Keyser-Dedekind-Hilbert species, (B) is a Calculus of Tensors and (y) is a Calculus of Matrices. In Part B on The Indeterminacy Principle Quantum Mechanical Equation in this our *Prolegomena*'s Volume I, it will be shown that this is precisely the Tripartite Calculus which the Einsteinean-Heisenbergean Epistemology (EEFP-III) and (EEFP-IV) and (HEFP-V) entails from which, as stated in our Sections S11 and S12, the Democritean-Eudoxean *Epistemological First Principles* (DEFP-I) and (DXFP-II) derive as special restricted cases.

Since 1970, Professor Fonseca and the present writer have combined forces, the present writer addressing the academic heads of the Humanities, Medical, and Mathematical Physical Departments in the University of Lisbon in 1975; Professor Fonseca reporting, a year later, on the crucial experimental verifications by himself and others, at the fall term research seminar held in the Law School of Boston

---

[59] Kenneth Craik, *The Nature of Explanation* (1943), Cambridge University.

College in October of 1976. Since then, more and more *Maximal Logical Consequents Deduced Theorems* have been put to crucial experiment test with positive results and since then many more such deduced theorems. It is the great merit of Trinity College Cambridge's late Professor Kenneth Craik to have seen as early as 1943 and later that only a relational entitied theory can account for the knower's knowing of the objects known.

Professor Fonseca's last letter of date October 28, 1982, reports as follows:

> As you may imagine, I am in an almost continuous "rush" in all senses — every bit of experimentation implies a very high amount of investment in experimentation in the laboratory but also a lot of deep epistemic analysis and finally a terrific fight to try to polish it from my own errors and those which form the responsibility of printers who do their work without knowing a word of English.
>
> I am excited now because a crucial neurophysiological experiment will be possible within two months, and, I think, a political, will be tested.

The flavor of the import of this achievement for the humanities, as well as for psychology, jurisprudence and especially aesthetics, may be expressed best by the following poem by today's Portuguese poet with which Professor Fonseca inscribed his 1970 book.

> To be great, be complete
> Nothing yours exaggerate or exclude
> Be all yourself in everything
> So in the lake, the eternal moon shines
> So high it lives.
>> Fernando da Sena

With these achievements, the rich implications of the Einsteinian-Heisenbergian *Epistemological First Principles* for the humanities seem secure. For without this scientific psychology, the connecting bridge between the verified exact natural science and each and every one of the humanities would not be. Our previous section of this our present *Prolegomena* has demonstrated also that the present exact sciences must not merely entail a richer humanities for tomorrow's world but must subsume within themselves as special restricted cases the verified 'EN APXAI of the Greek Euclidean and Latin Neoplatonic past, "Nothing yours exaggerate or exclude" to "be complete." Hence, our next section.

S14

The Geographical Locus and Historical Context of Greek-Euclidean Scientific and Greek-into-Latin Neoplatonic Humanistic 'EN APXAI, Including the Latin of Newton's *Principia*

Our quotation from Einstein in our Section S2 above, on the import of Euclid's geometry for the understanding of today's physics, a judgment in which Whitehead concurs, has been noted. Hence, the need to make the Greek *'EN APXAI* of ancient Greek verified mathematical physical natural science and its later Greek-into-Latin Neoplatonic era basic to our entire inquiry as given to us in the Heiberg Greek texts: *Axioms, Definitions, Postulates and Common Notions* of Euclid's (300 B.C.) *Elements*.

For the overall cognitive concerns of our present *Prolegomena*'s inquiry in the humanities as well as in the verified theoretical mathematical sciences in their architectonic diversified unity, we must supplement the two books by Professor B. L. van der Waerden, namely, his *Erwachende Wissenschaft: Ägyptische, Babylonische, Griechische* (1966) and his *Die Pythagoreer* (1979), from which we quoted so extensively in Section S11 above on Professor Burkert's classical paper on Democritos (1977), with three slightly earlier books by the former experimental quantum theory spectroscopist, Professor Samuel Sambursky, who today is the Director of the Institute on the History and Philosophy of Science in Israel's University of Jerusalem. These three books, which henceforth we shall denote by their dates are *The Physical World of the Greeks* (1956), *The Physics of the Stoics* (1959) and *The Physical World of Late Antiquity* (1962), all three published in English by Routledge and Kegan Paul Ltd. in London.

Let us begin with Professor van der Waerden's (1966) book first. With respect to the pivotal import of Euclid, he writes:

> Euclid collected the whole of mathematical physics known at the time of Plato (425-347 B.C.). All of us classical and modern mathematicians are his pupils. He and his Alexandrian successors (300 B.C.–140 A.D.) initiated the, if possible, more brilliant theorems of Archimedes, Eratosthenes and Apollonios . . . from which Kepler proceeded and upon which Newton built. (page 200)

On his page 331, as translated into English by the present writer, Professor van der Waerden continues:

With the insight of a leader guided by fundamental knowledge of the geographical context and of the future possibilities for commercial exchange and development, the young Alexander the Great determined upon Alexandria as its location and allowed the city to be built within a short time; Alexandria became a booming city and at the same time a cultural center of topmost quality. The kings Ptolemaios I Soter, Ptolemaios II Philadelphos and Ptolemaios III Euergetes, who ruled one after the other from 305 to 222 B.C. not only founded a mighty realm, but in a truly regal manner made possible both the arts and science. The first Ptolemy constructed the Museum, the Temple of the Muses, which brought together poets and learned scholars who were supported in a magnificent way with stipends from the Royal Treasury. In the Museum there was added a world famous Library to which Ptolemaios III Euergetes presented the entire Library of Aristotle and of Theophrastos; the most famous men of the then existing world of letters and the several sciences were together in Alexandria: Philologists (who called themselves Grammarians), Historians, Geographers, Mathematicians, Philosophers and Poets.

What is remarkable about this Macedonian Alexandrian Egyptian Greek epoch of the Ptolemies (which extends from 323 B.C. to 140 A.D.) is that, as Professor van der Waerden demonstrates in pages 176 through 330 of his treatise (1966), so far as irreducible 'EN APXAI First Principles are concerned, all of them were discovered, constructed, relational analytic R entitiedly defined, and crucially experimentally verified by the time of the completion of Euclid's Elements in 300 B.C.

For this reason, he calls, most aptly, this historical period "Das goldene Zeitalter" — "The Golden Age of Exact Greek Science." Roughly speaking, Professor Sambursky's first book, entitled The Physical World of the Greeks (1956), corresponds to Professor van der Waerden's "Golden Age of Greek Science." His second book is entitled The Physics of the Stoics (1959), because it falls into the Alexandrian post-Euclidean historical period; we shall call this period the Macedonian Ptolemaic Egyptian Greek era, with the dates 300 B.C. to 140 A.D., the greatest of the Stoics being the remarkable mathematical logician and theoretical mathematical physicist, Chrysippos, who lived between c. 280 B.C. and 207 B.C.

Of its humanistic impact and high quality, Professor van der Waerden in his 1966 work adds:

The Works of Homer were critically edited and freed from all interpolation, thereby grounding the science of Chronology, and the art of the poet was developed and refined. In Astronomy exact observations were made and Theorems were constructed, such as the Epicycle and

Eccentric, and the observations were explained in terms of Universal First Principles. The culmination of this Alexandrian development was the great *Syntaxis Mathematica*, the *Almagest* of Ptolemaios Klaudius in 140 A.D.; but the foundation of all these theorems were laid out at the beginning of his Epoch in Euclid's Greek *Elements* in Alexandria in 300 B.C.

Professor Van der Waerden's first sentence directly above reminds us that the science of historical scientific textual criticism of either secular or Holy Scriptural Greek texts did not begin, as most contemporary Biblical textual scholars suppose, in Pre-World War I's Germany but arose in Alexandria's Research Library in Euclid's Egypt in 300 B.C. and persisted there through the post-Euclidean Alexandrian Egyptian Research Library's influence to at least 140 A.D. It is from the *Syntaxis Mathematica* and *Nova Astronomica* of Ptolemaios VI Klaudios in 140 A.D. that Kepler's treatise with his Three Laws of Planetary Kinematic Motions arose and without which also Newton's Latin *Philosophiae Naturalis Principia Mathematica* (1686) would not be. Meantime, also, the Physics, Metaphysics and Theology of Philo of Alexandria (25 B.C.–22 A.D.) spread (after the two previous Babylonian Conquests of Palestine and the entire Middle East) in the Post Philo Judaic later prodigiously influential cultural "Dispersions". Without this influence also from the Research Egyptian Library of the Alexandrian Macedonian Septuagint Greek dialect Ptolemies, in this Alexandrian Greek era between 323 B.C. and 140 A.D., Jewish culture in addition would not be, nor Judaic-Greek Christian either.

Thus it is that Sir Lancelot Lee Brenton, in his indispensable Introduction (pages i-vi) of his edition of *The Septuagint Version of the Old Testament*, writes as follows:

At Alexandria (1) the Hellenic Jews used this version, and gradually attached to it the greatest possible authority: from Alexandria it spread among (2) the Jews of the Dispersion, so that by the time of Our Lord's birth (also that of Philo of Alexandria's (25 B.C.–22 A.D.) *Physica, Metaphysica* and *Theologica*) it was the common form in which the Old Testament scriptures had been diffused.

On the next page (iv) of his Introduction, Sir Lancelot continues:

After the diffusion of Christianity, copies of the *Septuagint* became widely dispersed through the new communities that were formed so that before many years had elapsed, this version must have been in the hands of Gentiles and of Jews.

Most exciting of all is Sir Lancelot's account of the circumstances under which, from a single perishable parchment text (analogous to that of the few fragments of the much overworked present-day Dead Sea Aramaic Scrolls), the translation in Septuagint was made and his account of the cultural linguistic and commercial background of the Jews who made this translation, and, also, when and where they made it from the perishable parchment Hebrew text into the Septuagint Greek text. On all these exceedingly important yet for the most part rarely mentioned in our present day points, Sir Lancelot writes as follows:

> The earliest version of the Old Testament which is extant, or of which we have certain knowledge, is the translation executed at Alexandria in the third century (B.C.) before the Christian era. This version has been so habitually known as the *Septuagint* that the attempt of some learned men . . . to introduce the designation of *The Alexandrian Version* (as more correct) has been far from successful.

Then, after describing and demonstrating that certain reports and their pseudo-versions are erroneous, Sir Lancelot adds:

> The fact may, however, be regarded as certain, that prior to the year 285 B.C. the Septuagint version had been commenced, and that in the reign of Ptolemaios II Philadelphos, either the books in general or at least most of them had been completed.

So much for the temporal period during which the translation from a Hebrew papyrus manuscript, the 70 Books, "or at least most of them," were translated in Alexandria into Septuagint Greek. What about (1) the Hebrew manuscript itself from which this translation into Septuagint Greek was made and (2) the Jews who made the translation? With respect to the first (1) part of this key question, Sir Lancelot replies that:

> It was an Egyptian King who caused this translation to be made and it was from the Royal Library at Alexandria that the Hellenistic Jews received the (Hebrew) copies which they used.

The adjective *"Hellenistic"* in this answer to our query (1) provides the answer also to its part (2). This becomes evident when Sir Lancelot, appealing to internal Septuagint Greek criteria of scientific authenticity (recall what Professor van der Waerden has demonstrated with respect to the critical textual analysis of Homer), answers part (2) of our question thus:

> In examining the reason itself, it bears manifest proof that it was not executed by Jews of Palestine, but by Jews of Egypt.

To understand this surprising fact, two factors must be known. The first one is that the Jews of Egypt were refugees (in the Nile delta with its entire international commercial trade) from the two previous Babylonian captivities. For the second and key factor, we are again indebted, as usual, to Professor van der Waerden. In his *Erwachende Wissenschaft: Ägyptische, Babylonische, Griechische*, he tells us that:

> For six generations antecedent to the time of Euclid (300 B.C.) Athens enjoyed a monopoly on all the international trade with the Egyptian Nile Valley centered in Alexandria.

When, therefore, Alexander the Great founded the great Library in Alexandria in 323 B.C. under his general Ptolemaios I Soter, the Jewish refugees in Alexandria, being "in trade", found it necessary to learn the Macedonian Septuagint Greek dialect rather than classical Athenian Greek.

Consequently, when the Egyptian Greek translators of the Septuagint Old Testament were given by the Ptolemies I Soter and II Philadelphos the fragile papyrus Hebrew manuscript, they were using their own "six-generations-old" Septuagint Greek language in making their translation in the Research Library in Alexandria, between, as Sir Lancelot Lee Brenton tells us, 323 B.C. and 285 B.C. Hence, Sir Lancelot's demonstration that this translation was made "not by Palestine Jews, but by Septuagint Greek speaking, thinking and writing Jews", presents no problem by way of explanation whatever.

Even more important, and (to the present writer's knowledge) noticed for the first time here, at the very time when these "six-generations-old" Alexandrian Macedonian Greek-minded Egyptian Jews were translating the fragile Hebrew parchment manuscripts given them by Ptolemaios I Soter into their and his natural Septuagint Greek language in the great Library in Alexandria, Euclid had in 300 B.C. completed there the *Thirteen Greek Books* with their three Greek defined 'EN APXAI and was lecturing there upon them. Under these circumstances, we find it well-nigh impossible to suppose that the Egyptian Jewish translators were not aware of these 'EN APXAI, especially the feminine gendered 'Εν ἀρχῇ one that is defined in Verses 1 and 2 of Genesis, and used them in their Septuagint Greek translations, thereby in their minds, as in the case of Philo (c. 30 B.C.–45 A.D.) of Alexandria's *Physica, Metaphysica* and *Theologica*. Undoubtedly also this gave not Philo of Alexandria, but also all the Jews of the subsequent Dispersion, the conviction that the 'EN APXAI of Philo's *Physica*, being crucial experimentally in Alexandria's Euclidean Greek *Physica,* holds for its *Metaphysica* and *Theologica* also.

In any event, by way of summary of the contributions by Professors Van der Waerden and Sambursky, and by Sir Lancelot Lee Brenton, we arrive in this our Section S14 with neither Palestine nor Athens as the geographical locus of Hebrew-into-Septuagint Greek verified scientific origins but instead at the Research Library in the first two Ptolemies' Alexandria between 323 and 290 B.C. when Septuagint Greek dialect Jews were translating the seventy books of their Holy Scripture from a fragile parchment Hebrew text into Septuagint Greek at the time that the great Greek Euclid had completed his 'EN APXAI or first principles, had set them down in his *Thirteen Greek Books or Stoicheia* and was lecturing on them.

We have now but to bring together the findings of Professors B. L. Van der Waerden and Samuel Sambursky, as described just above, to arrive at the three following historic periods which cover the Hebrew-into-Septuagint Greek-Greek Euclidean Ptolemies, the crucial experimentally verified Euclidean *'EN APXAI* scientific and the Greek and Greek-into-Latin Neoplatonic historical parts of this our present *Prolegomena*'s overall inquiry's theme. The three historical periods are:

    I. *The Golden Age of Greek Science (610 B.C.–300 B.C.)*
    II. *The Macedonian Egyptian Era of the Alexandrian Ptolemies (325 B.C.–140 A.D.)*
    III. *The Neoplatonic Era of Late Antiquity (203 A.D.–650 A.D.)*

It might seem that at long last our *Prolegomena* has reached its end. However, one variable, the final word in our *Prolegomena*'s S1 title, namely, "Christianity," still remains to be given a rigorous relational analytic *R* entitled definition.

S15

Hebrew-into-Septuagint Greek-Greek New Testament Christianity Has Its Greek 'EN APXAI Definition Also

The first two are so obvious, and the third one is so frequently used, and even prayed, that one wonders why all three parts of this Greek definition have not been noted, thought and preached in every synagogue and Protestant and Roman Catholic church and even secular school today.

The first one (a) is the feminine gendered Ἐν ἀρχη which is negatively defined identically in both Verses 1 and 2 of the Hebrew-into-Septuagint Greek Genesis and in Definitions 6 through 1 in Book I, and more especially in Proposition 20, Line 26 in Book III of Euclid's Greek *Elements*. The second one (b) is the positively defined masculine gendered 'EN APXH of St. John's Verse 1 in the Fourth Gospel. The third, which is not so obvious, was called to the present writer's attention by the Reverend Malcolm Matheson (Harvard Divinity School B.D.), when he referred to the following statement made by the great Tertullian in his "An Interpretation of the Lord's Prayer." As translated into English, Tertullian's statement is: "*The essence* [i.e., the definition] *of Christianity is in our Lord's Prayer.*"

For this naturalistic *is,* as distinct from the obligational jurisprudential ought portion of *The Lord's Prayer,* this gives us (c) the positively defined masculine gendered *'EN APXH,* as defined in the *first Greek line* of *Our Lord's Prayer,* as given in *Line 9* of *Chapter 6* in *St. Matthew's Gospel.* Then, combining (a), (b) and (c) our tripartite definition of Hebrew-into-Septuagint Greek-Greek New Testament Christianity becomes:

(a) The feminine gendered Ἐν ἀρχη of Hebrew-into-Septuagint Greek Genesis, as defined in its Verses 1 and 2.

(b) The Trinitarian thrice-repeated λόγος ὁ θεός masculine gendered *'EN APXH,* as defined in Verse 1 of St. John's Gospel.

(c) The πάτερ ἡμῶν, masculine gendered 'EN APXH, as defined relational analytic R-wise ἐν τοῖς οὐρανοῖς pluralistically by these last three Greek words in the first Greek line of *Our Lord's Prayer* in Chapter VI of ΚΑΤΑ ΜΑΘΘΑΙΟΝ.

Ergo, QED: the tripartite definition of Hebrew-into-Septuagint Greek-Greek New Testament Christianity.

No one should attempt any English or any other ordinary language translation of the three Greek *'EN APXAI* above or any other text in the *Old* or *New Testament* until he can define (a), (b) and (c) in Greek, as those three (a), (b) and (c) definitions are given in a trustworthy Greek text. This forces us to our next and last Section's topic.

S16
The Greek Textual and Grammatical Scriptural Tools for Our Task

For our definition of (a), our feminine gendered Ἐν ἀρχῆ will be the *Septuagint* version of the Old Testament as edited by Sir Lancelot Lee

Brenton. For our masculine gendered definition of the *'EN APXH* defined in (b) and (c), our Greek New Testament text not merely will, but also *must*, be the New Testament in the Original Greek, the text revised by Brooke Foss Westcott and Fenton John Anthony Hort (edition 1914 or earlier). Why this "also must be"?

The answer is a very simple but tragically serious one. Believe it or not, the recent Common Bible Committee, of which the late Dean Luther A. Weigle was the Chairman, and who in 1916 was the present writer's teacher on the Christian Nurture of Horace Bushnell, defines the ὁ λόγος of our masculine gendered *'EN APXH* in Verse 1 of St. John's Gospel as opening with the two Greek words Ἐν ἀρχῆ and therefore as being (all Greek nouns ending in α or ῆ being) feminine gendered. Hence, what occurred is that the Jewish, British, Scottish, Vatican Roman Catholic, Greek Eastern Orthodox and the American members of this Common Bible Committee *did* what Dean Inge and Spengler warned against in 1923 when, as quoted by us in the introductory section S1 of this our *Prolegomena,* Dean Inge predicted that

> ... those who would follow Luther and Harnach's inclination to by-pass Greek Christian civilization for nothing but its Hebrew component risk destroying Christianity.

The German Spengler added in his *Untergang des Abendlandes* that the same holds true for Western natural science and civilization generally.

Little did we realize on page 1 of this *Prolegomena* that but a few pages later both the then Dean of St. Paul's London and Spengler's predictions, the latter made in Munich, Germany, would have come true. In any event, it is on this tragic note that this *Prolegomena* ends and that the task of its four future volumes becomes that of turning this sequence of historical events around.

Since our beloved friend and tutor Albert Einstein suffered from faulty or omitted translations and mis-translations, as has Holy Scripture, it is appropriate to close this our hard-won *Prolegomena* with the following positive and optimistic words as quoted from him in the September 1981 issue of the *AAAS* journal *Science.*

> I want to know how God created the universe. I am not interested in this or that spectral line or other phenomenon. I want to know God's mind. The rest are details.

FINIS PROLEGOMENA

BIBLIOGRAPHY

I. *Original Greek and Greek-into-Latin 'EN APXAI Texts*

Heath, T.L. *The Thirteen Books of Euclid's Elements,* translated from the Heiberg Text (300 B.C.) with Introduction and Commentary (1908). Volumes I, II, III and IV. Hereafter (H-1), (H-II), and (H-III) with Index of Greek Words and Forms (H-I, pages 411-412), (H-II, pages 427-429), and (H-III, pages 529-534). Always in English translations Heath's Heiberg Greek Definitions and their *Indices of Greek Words and Forms* are to be used.

Diels and Others. *Greek Fragments.*

Bakewell, Charles Montague. *Source Book of Ancient Philosophy* (1907). Charles Scribner's Sons, New York, Chicago, Boston. Check all English translations with those of Heath's (1908) Greek Words and Forms in (H-I) (H-II) and (H-III).

Quellen und Studien zur Geschichte der Mathematik, Astronomie und Physik. Quellen A–Studien B. (1932) Verlag von Julius Springer, Berlin

Luria, S., *Studien B. "Die Infinitesimale Theorie der antiken Atomisten."* Vol. 2., pages 107-185.

Toeplitz, Otto. *Das Verhältnis der Mathematik und Ideenlehre bei Plato Bol. Ibid.,* pages 12-33.

Julius Stengel, *ibid.*, Studien B. *"Zur Theorie der Lógos bei Aristoteles.* Pages 34-66.

Frank, Erich. *Plato und die Sogenannten Pythagoreer* (1923). Verlag von Max Niemeyer Halle (Savle). Pages 1-399.

Sachs, Eva. *"Die fünf platonischen Körper. Philolog. Untersuchungen* (1917). Heit 24, pages 1-242.

Van der Waerden, B.L. *Die Pythagoreer: Religiöse Brüderschaft und Schule der Wissenschaft* (1979). Pages 1-505. Artemis Verläg Zürich und Münich.

Burkert, Walter. *Air-Imprints of Eidola: Democritos' Aetiology of Vision* (1977). *Illinois Classical Studies,* Vol. II. Miroslav Marcovich, editor, pages 97-109, University of Illinois Press.

Plato. *Opera:* tomi I-V, ed. Ioannes Burnet. Oxonii e Typographeo Clarendonianó, 1899-1906.

Aristotle ΑΡΙΣΤΟΤΕΛΗΣ. *Historia Animalium, De Partibus Animalium, Physica, De Calo, De Generatione et Corruptione* 325 25, *Metaphysica, De Anima,* should be read in this order and related to Eudoxos' Proof of Prop. I in Euclid's *Book X* and to Luria's article above.

Northrop, F.S.C. "The Mathematical Background and Content of Greek Philosophy." Pages 1-40 in *Philosophical Essays for Alfred North Whitehead* (1936). This correlates all the ancient Greek Scientists, Philosophers and Humanists from Anaximander (610 B.C.) through Parmenides, Anaxagoras, Democritos, Archytas, Plato, Eudoxos, Aristotle, and the Stoics with the respective Thirteen Greek Books of Euclid's *Elements* in which the respective *First Principles* (Greek *'EN APXAI*) of each are defined with their respective deduced theorems proved.

van der Waerden, B.L. *Die Pythagoreer* (1979). Artemis Press. Zürich und Munich. *Erwachende Wissenschaft: Ägyptische, Babylonische, Griechische.* Zwe Leige Erganze Verlag (1966) Birkhäuser Verlag. Basle and Stuttgart.

Sambursky, Samuel. *The Physical World of the Greeks* (1956); *The Physics of the Stoics* (1959); *The Physical World of Late Antiquity* (1962). Routledge and Kegan Paul Ltd. London.

The *Isagoge of Prophesy* on *Aristotle's Logic of Predictables.* A translation by Charles Glenn Wallis based on that of Octavius Fraire Owen M.A.

Morrow, Dwight R. *Proclus. A Commentary on the First Book of Euclid's Elements* with *Introduction* (xv-xiv). English translation, pages 1-355. Princeton University Press (1970). Princeton, New Jersey.

II. *The Relation of the Above Greek-Latin 'EN APXAI to Today's Exact Verified Science*

Philo of Alexandria (c. 30 B.C.-54 A.D.) The *'EN APXAI* common denominator in *Euclid's Book I* and the *Hebrew-into-Septuagint Greek Genesis, Verses 1* and *2*; as given in *De Morgan's Questiones et Solutiones.* For the relation to today's *First Principles,* see S. Sambursky, *Physical Thought from the Presocratics to Quantum Physicists* (1975), Pica Press, N.Y.

Brenton, Sir Lancelot Lee. *The Septuagint Version of the Old Testament* with *Introduction* (i-iv). Hebrew-into-Septuagint Greek Text, pages 1-1139. Greek Text Column of Each Page. Samuel Bagstar and Sons. London. (No Date). The Introduction is interesting reading.

Tertullian (160 A.D.-240 A.D.). The Greek-into-Latin translation of Holy Scripture including the Latin of Newton's *Philosophiae Naturalis Principia Mathematica*.

Ptolemaios Klaudios (140 A.D.). *Syntaxis Mathematica*, and *Nova Astronomica*. Kepler and Galilei, and Newton inherited the verified *'EN APXAI* of ancient Greek science by way of these two Greek Treatises and also Apollonius of Perga's Treatise on Conic Sections. (See van der Waerden (1966) *op. cit.*)

St. Thomas Aquinas. *A Commentary on Aristotle's Physics* (1245 A.D.). English translation, pages 1-595. *Summa Theologica* (1274 A.D.). Roulledge and Kegan Paul (1963) London.

III. *The Transmission of the Verified Ancient Greek 'EN APXAI to Today's Verified More General First Principles Science*

Kepler, Johann. (1571-1630). *Astronomia Nova ά ιτιολογητοσ, Seu Physica Coelestis*, Prague, 1609. Gives the Laws of Elliptical Orbits and Equal Areas.

Galileo Galilei (1564-1642). *Dialogo dei Massima Systemi dei Mundo* (1630), *Dialogo delle Nuove SC* (1636).

Sambursky, Samuel. "Three Aspects of the Historical Significance of Galileo," in which Galileo Galilei quotes Tertullian (190 A.D.) with approval on the relation between science and the Christian religion, in Volume II, No. 1, page 3 *Galilei Opere V, p. 316* of the *Proceedings* of *The Israeli Academy of Sciences and the Humanities*. Jerusalem 1964.

Newton, Sir Isaac. (1686). *Philosophiae Naturalis Principia Mathematica. Volumes I and II* (the Third Edition [1726]) with *Variant Readings Assembled and Edited* by *Alexandre Koyré* and *I. Bernard Cohen*, with the assistance of Anne Whitman. Harvard University Press (1972). Copyright by the President and Fellows of Harvard College. Typeset at the Printing Press, Cambridge, England. English translation by Andrew

Molte in 1729. The translation revised and supplied with an
historical and explanatory appendix by Florian Cajori, late
Professor of the History of Mathematics Emeritus at the
University of California, Berkeley, 1934. The above Koyré
and I. Bernard Cohen *Volumes I* and *II* edition of 1972 has
an excellent up to date (1978) Independent Introduction by
I. Bernard Cohen, pages 1-380, to Newton's "Principia"
(1978). Harvard University Press. Cambridge, Mass. U.S.A.
Faraday, Michael (1791-1867), and Maxwell, James Clerk
(1831-1879). *Experimental and Rigorous Theoretical Mathematical Physical Equational* (1841) and *Field Physical Verified
Theory of Electricity and Magnetism* (1873).

IV. *The Alternative Theories and Their Respective Equations of the Revolu-
tion in the First Principle of the Sciences and the Humanities Entailed
by the Incompatibility of Newton's Particle Physics (1686) and the Faraday Maxwellian Field Physics (1841)*

Whitehead, Alfred North. *The Mathematical Concepts of the Mate-
rial World. Philos. Transactions Royal Soc. of London; Series A.
Vol. 205 (1906). The Axioms of Descriptive Geometry. Cambridge
Tracts in Mathematics and Mathematical Physics*. No. 5. VIII.
Pages 1-74. Whitehead and Russell *Principia Mathematica.
Vol. 1 (1910), Vol. II (1912), Vol. III (1913)*. Cambridge-at-
the-University Press. England, "*Space, Time, and Relativity*".
Proc. of the Aristotelian Soc., Vol. 16, pages 104-129. *An Inquiry Concerning the Principles of Natural Knowledge* (1919)
Part III Whitehead's epistemological "*Method of Extensive
Abstraction*". *The Concept of Nature* (1920); Whitehead shifts
from its previous "neutral monistic" theory of the self, and
his *Chapter* on *Congruence* demonstrates that Einstein's Met-
rical heterogeneity leaves astronomical location of distant
stars impossible (van der Waerden later makes this same
point). Cambridge-at-the-University Press. *The Principle of
Relativity, With Applications to Physical Science* (1922). CUP.
"*Uniformity and Contingency*". *Pres. Inaugural Address. Proc.
Aristotelian Society* (1922-1923). *n.s.* Vol. 23, Pages 1-18.
"*Science and the Modern World*". (Lowell Lectures) (1925).
Macmillan, N.Y. *Religion in the Making* (1926). Macmillan,

N.Y. *Process and Reality* (Gifford Lectures). Macmillan, N.Y. (1929). Pages 1-547. *Immortality.* The Ingersoll Lecture for 1941. Harvard Memorial Church, April 22, 1941.

Einstein, Albert. *Albert Einstein's "First" Paper: "Concentrating the Investigation of the State of Aether in Magnetic Fields" with a letter to Cäsar Koch* (1894 or 1895), as described by *Jagdish Mehra* in *Science Today* (April 1971).

"Zur Elektrdynamik bewegter Körper, Annalen der Physik" Band 17 (1905)

"Zur Theorie der Brownischen Bewegung." Annalen der Physik Band 19 (769-822) Grundlage der allgemeinen Relativitätstheorie. Band 49 Seite 768-822; Epistemological Principles in German, pages 769-776; Deduced (1) - (4) - (75) Equations 777-882.

"Quantentheorie der Strahlung" (1916) *Physikalische Gesellschaft Zürich Mitteilung Vol. 16* (47-62).

H. Minkowskei (1908) "Raum and Zeit". Address given to Association of Natural Scientists and Physicians 21 Sept. 1908 Cologne Germany.

Walter Thirring (1972). *Remarkable Logical Universality,* and *Verification by Others of Einstein's General Equation".* Vol. 4, *Essay in Physics,* Chapter 4 in Cohn and Fowler (1972) *Essays in Physics.* Academic Press, London and New York.

Albert Einstein. *"Auf die Riemanns — Metrik und den Fern-Parallelismus begründete Einheits — Feldtheorie", Mathematische Annalen,* Vol. 102 (1930), pages 685-697.

V. *Original Language Papers of Indeterminacy Principle Quantum Mechanics*

Planck, Max. *Zur Theorie des Gesetzes der Energieverteilung Normal Spectrum.* Verband der Deutschen Gesellschaft (1900). Pages 237-245.

Einstein, Albert. *Zur Quantentheorie der Strahlung. Physik Annalen Band 18 (1917). Seite 121* ff. Eng. trans. in B. L. van der Waerden, *Sources of Quantum Mechanics.* Pages 63-77. Hereafter (WSQM) P. B. (1968) Dover Pub. Inc. N.Y.

Rutherford, Sir Ernest (1911). *100% Deterministic Mechanical Causal Model of the Dynamic Atom.* Experiments in Cavendish Laboratory (1911-1916).

Bohr, Niels. *On the Quantum Theory of Line Spectra* (April 1918) in *Danske V. d. Selak Nat-Math. Afd. 8 Raekke IV. 1.* Eng. trans. in *B. L. van der Waerden, Sources of Quantum Mechanics.* Pages 95-138.

Ehrenfest, P. *"Adiabatic Invariants and the Theory of Quanta." Phil. Mag.* 33 (1917). (WSQM). Pages 79-93.

Schrödinger, Ervin (1922). *Annalen der Physik Band 79* (1926). Pages 361 ff.

Heisenberg, Werner (July 29, 1925). *"Über Quantentheoretische Andeutung kinematischer und mechanischer Bezeichnungen". Zeitschrift Physik. Band 33 (July 29, 1925).* Concerning Quantum Theoretical Significance of Kinematic and Mechanical Relations." This is his "Observables Only" paper (WSQM). Pages 261-276.

Pauli, Wolfgang. *"Über das Wasserstoffspektrum vom Standpunkt der Neuen Quanten mechanik. Zeit, Phys. 36 Jan. 17* 1926.

Dirac, P. A. M. *"The Fundamental Equations of Quantum Mechanics." Proc. Royal Soc. A. Vol. 109* (Nov. 7, 1925). Pages 642 ff.

Heisenberg, Werner. *The Physical Principles of the Quantum Theory* (1930), University of Chicago Press. Pages 1-186.

Schrödinger, Ervin. P. A. M. Dirac and Infeld and van der Waerden (1931). *Three Independent Proofs that the (1900-1927) Indeterminacy Principle Quantum Mechanical Equation Satisfies Einstein's Two Epistemological First Principles Requirements for Scientific Objectivity — (EEFP-III) and (EEFP-IV).*

Heisenberg, Werner. *The Principle of Limited Logical Universal Domains [HEFP-IV].* In *Dialectica* (1948), Neuchatel, Switzerland.

VI. *Towards the Discovering and Construction of a Verified Intensional n = 3 adic Neuropsychologically Embodied Logic of Psyche*

von Domarus, Eilhard, M.D. Natural History Psychiatry Whiteheadean Epistemologically Based Ph.D. in Philosophy at Yale University on the Logical Structure Mind (1934). Published as Chapter XV in *Communication: Theory and Research.* Lee Thayer, editor. Charles C Thomas, Publisher (1967).

McCulloch, Warren S. Chapter XIV and XV in the Symposium Volume *Communication: Theory and Research*, Lee Thayer editor, Charles C Thomas, publisher, Springfield, Ill., 1967. "Being a Belated Introduction to the Thesis of Eilhard von Domarus," and in *Chapter XVI* of *the direction we must take, neurocybernetically and psychologically Embodied in a manner yet to be discovered* that satisfies Einsteinean Epistemology First Principles (EEFP-III) and (EEFP-IV) and also Heisenberg's (HEFP-V).

da Fonseca, J. Simões, M.D. in Clinical Psychiatry at the Medical School of the University of Freiburg, Germany (cf. von Domarus above) and Director Ph.D. of the Neuropsychological Laboratoria de Faculdade de Medicina de Lisboa, Portugal. *Signification and Intention.* (April 1970) *Ch. I. Psychological Processes from the Viewpoint of Signification and Intention.* Pages 1-4, *A Calculus of Relations (*α*)* the Operational Definitions for Verifications by himself and Others being *(*β*) a Calculus of Tensors,* and (γ) *a Calculus of Matrices.* Publishers: *Faculdade de Medicina de Lisboa Laboratorio Neurofisiologia. Centro de Estudo Egas Moniz.* Pages 1-97.

VII. *The Diverse Logical Instruments Required*

(NHSI). Aristotle's Natural History Clinical Data *Logic of Predictables.*

(ELC). The Extensional Logical Calculus of *Whitehead and Russell's Principia Mathematica's* $\phi(x)$, $\supset$, $\psi(x)$ single entity-property *Theory of Deduction.*

(THS). The Theoretical Mathematical Science of a Statics only.

(KBW). The Kronecker-Brouwer-Weyl Whole $\alpha\gamma\alpha\lambda\lambda o\gamma o\nu$ Theory of Pure Mathematical and hence, *ipso facto,* of Mathematical Physical Entities also.

(DWFF). The Dedekind-Weierstrass-Hilbert theory of "der weniger welcher: Funktion Funktionen — "the one and the only one — which" definite-articled relational theory of pure mathematical entities, which is not merely *ipso facto* a theory of mathematical physical entities, but also achieves this with no "set theoretical paradoxes," because a function can take itself as the value of its own variables.

VIII.   *Necessary Original Texts on the Humanistic Implication of the Verified 'EN APXAI Texts I through VII above*

Brenton, Sir Launcelot Lee. *The Hebrew-into-Septuagint Greek Old Testament. Genesis Verses 1 and 2.* Bagster and Sons Ltd. London, n.d.

Das Alte Testament (1966). *Ausgewählt, Übertragen und ingeschichtlicher Folge angeordnet*, The Old Testament, edited, critically revised and ordered in historical sequence by Jörg Zink (1966). Kreuz-Verlag. Stuttgart-Berlin.

Schoonenberg, Piet. *Bund und Schöpfung (Covenant and Creation).* (1970). *Genesis Chapter I-II. Verse 4.* Benziger Verlag. Zurich, Einsiedeln. Cologne.

The New Testament in the Original Greek. The text revised by Brooke Foss Westcott, D. D. and Fenton John Anthony Hort, D. D. Macmillan and Co., Ltd. St. Martin's Street, London (1914).

Nunn, The Reverend H. P. V. *The Elements of New Testament Greek.* Edition 1955 or earlier. Cambridge at the University Press. England.

Novum Testamentum et Psalterium (1974). *Typis Polyglottis Vaticanis Iuxta Novae Vulgatae Editionis Textum cum Indice analytico-alphabetico et Appendice Precum.*

Inge, William Ralph. *Gifford Lectures* (1917, 1918), *Vols. I and II, Philosophy of Plotinus (Enneads).* Published (1923) Longman's, Green, Co. London, New York, Toronto, Bombay, Calcutta, and Madras.

Index